SPATIAL INFORMATION TECHNOLOGY
FOR
NATURAL RESOURCE MANAGEMENT

About the Editors

Noor Mohammad, (b. 1942), after obtaining his M.A. and Ph.D. degrees in Geography, joined Kirori Mal College in 1969. Since 1976 he is associated with the Delhi School of Economics, University of Delhi, as Lecturer, Reader, Professor, Head of the Department and Co-ordinator UGC-SAP. Prof. Mohammad is actively engaged in research in the field of agricultural geography, landuse and planning and population geography. He has published more than 5 dozen research papers in many scholarly referred journals. His books *Agricultural Landuse in India* and *Caste and Primary Occupations: A Geographical Analysis* have been very well received and have won wide acclaim in Indian and Foreign universities. Prof. Mohammad has edited *Perspectives in Agricultural Geography* (in Five Volumes) and *New Dimensions in Agricultural Geography* (in Eight Volumes) under Concept's International Series in Geography which has been well taken the world over.

R.B. Singh is presently Reader and Dy. Coordinator-UGC SAP in the Department of Geography, Delhi School of Economics, University of Delhi. He has specialized in Environmental Studies, Disaster Management, Remote Sensing and GIS Technology. He has to his credit 26 research volumes/books and 145 research papers published in national and international journals. He is Secretary General, National Association of Geographers, India (NAGI) and is Steering Committee Member of the IGU Commission on LUCC since 1996. He is associated with IGU as Regional Promoter of South Asia for initiative on Culture and Civilizations for Human Development. He is also associated with four prestigious International collaborative research programmes such as ICSSR-IDPAD, CIDA-SICI, DFID and SHARP as Project Director.

Anindita Datta is a Senior Lecturer at the Department of Geography, Delhi School of Economics, University of Delhi. Her research interests are primarily within the field of social geography and gender and development issues. She has earned her doctorate from the Jawaharlal Nehru University, New Delhi on the theme of gender disparities in well-being. Dr. Datta has been associated with the Panjab University, Chandigarh and briefly with the Norwegian University for Science and Technology, Trondheim, Norway earlier in her career. Apart from publishing regularly in internationally known journals, she is part of the Department's University Grants Commission funded Special Assistance Programme.

Spatial Information Technology for Natural Resource Management

Editors

Prof. Noor Mohammad
Dr. R.B. Singh
Dr. Anindita Dutta

Published for
Department of Geography
Delhi University
by

CONCEPT PUBLISHING COMPANY, NEW DELHI-110059

Cataloging in Publication Data—DK
Courtesy: D.K. Agencies (P) Ltd. <docinfo@dkagencies.com>

Spatial information technology for natural resource management / editors, Noor Mohammad, R.B. Singh, Anindita Dutta.
p. cm.
Papers presented at the National Seminar on Spatial Information Technology and Natural Resource Management, held at New Delhi.
Includes bibliographical references.
Includes index.
ISBN 8180693589

1. Natural resources—Management—Geographic information systems—India—Congresses. 2. Natural resources—India— Management—Remote sensing—Congresses. I. Mohammad, Noor, 1942- II. Singh, R. B. (Ram Babu), 1955- III. Dutta, Anindita.

DDC 333.709 54 22

ISBN 81-8069-358-9

First Published 2007

Published and Printed by
Ashok Kumar Mittal
Concept Publishing Company
A/15-16, Commercial Block, Mohan Garden
New Delhi-110059 (INDIA)
Phones : 25351460, 25351794
Fax : 091-11-25357103
Email : publishing@conceptpub.com

PREFACE

The natural resources are continuously shrinking and deteriorating due to their excessive and unscientific utilization. Hence, there is an urgent need for the conservation and management of all kinds of natural resources. By and large, natural resource management of a region is performed either by manual map superimposition method or by computer technology using numerical spatial data sets or geo-based files. Manual superimposition of transparent maps with a view to produce multi-temporal maps has its limitations. Only a few maps can be superimposed simultaneously for visual analysis. The computer assisted approach requires information in digital form. Interrelation of resource features (attributes or factors) is performed through numerical transformations. The development to modern data-gathering techniques and computer-assisted cartography give new impetus to spatial data storage and exchange. Simultaneously, the combined use of Geographic Information System (GIS) and Digital Image Processing (DIP) provide better prospects of applying SIT for resource monitoring and forecasting over wider areas within limited time span. The Global Positioning System (GPS) provides three-dimensional positions 24-hour-per-day basis all around the world. This provides latitude, longitude and altitude of any spatial unit. Thus, the GPS has been contributing greatly in the field of spatial mapping. It would be easy to develop predictive modelling capabilities for effective public policies in years to come.

Considering the above issues, our Department under the Special Assistance Programme (SAP) organised a National Seminar on *Spatial Information Technology and Natural Resource Management.* The present book is an outcome of this seminar.

Although a large number of abstracts and articles were received and presented in this seminar, and they have been included in the Proceeding of Seminar, but only 19 articles of high academic standard written by eminent scholars have been published in the book. These articles have been coherently put into 5 sections according to their focal themes. Section I including four articles covers the conceptual aspects of spatial information technology, while Section II including four articles contains the papers dealing with the application of SIT in geomorphology, climatology and water resources. Four articles dealing with land use and land cover have been put in Section III, whereas the papers discussing agriculture and related fields have been placed in Section IV. The last section includes only three articles connected with forests and their management.

The department sincerely thanks University Grants Commission for sanctioning SAP with the provision of fund for holding the annual seminar. We are also thankful to Indian Council of Social Science Research (ICSSR) and the Council of Scientific and Industrial Research (CSIR) for extending generous financial support for the seminar. We are highly grateful to all the members of the SAP Advisory Committee specially Prof. L.S. Bhatt and Prof. (Mrs.) K. Sita (UGC Nominee) for their constant help and valuable suggestions for publishing the proceeding.

We also appreciate the sincere efforts and contributions made by Dr. Sanjay Aggarwal, Research Fellow for streamlining and producing the Proceeding of the seminar including abstracts and full papers. We would like to appreciate the constant efforts of Ms. Nidhi Gandhi and Ms. Ritika Shrimali, Research Fellows of the UGC-SAP in the publication of the book.

Editors

CONTENTS

SECTION III
LANDUSE AND LANDCOVERS

SECTION IV
AGRICULTURE RELATED FIELDS

SECTION V
FORESTS AND THEIR MANAGEMENT

INTRODUCTION

SPATIAL INFORMATION TECHNOLOGY FOR NATURAL RESOURCE MANAGEMENT

Geography is undergoing dramatic and substantial changes driven primarily due to technological developments in electronics, space and telecommunication fields. Geographers have a greater opportunity to utilize computer and multimedia for geospatial data processing. Every year a large number of new softwares are either made or modified. The Spatial Information Technology (SIT) has changed mapping in many ways. Softcopy images have replaced printed maps for many applications. Production time has been reduced drastically for micro planning and development needs through timely and high-resolution image of a designated site. Automation is also changing the quality as well as variety of spatial information that can be mapped. Further, Geographic Information System (GIS) is expanding the role of spatial mapping and analysis in decision making. Only few studies, based on such technology have been made in India in relation to natural resource management. In this sense, the SIT is emerging as pioneering technology of great significance and may serve as a powerful aid in the development planning and governance. To make the best use of the SIT, effective networking and multi-level infrastructure is essential. It bridges the gap between the users, end-users and data providers. It is important that sharing digitised spatial data as well as socio-economic and statistical data is a key issue to promote sustainable development. Therefore, spatial technology has become a key science for decision support of sustainable development and management of environmental resources.

SPATIAL INFORMATION TECHNOLOGY AND RESOURCE MANAGEMENT

Natural resource management of a region is performed either by manual map superimposition method or by computer technology using numerical spatial data sets or geo-based files. Manual of transparent maps with a view to produce multi-temporal maps has its limitations. Only a few maps can be superimposed simultaneously for visual analysis. The computer-assisted approach requires information in digital form. Interrelation of resource features (attributes or factors) is performed through numerical transformations. The development to modern data-gathering techniques and computer-assisted cartography give new impetus to spatial data storage and exchange. Simultaneously, the combined use of Geographic Information System (GIS), and Digital Image Processing (DIP) provide better prospects of applying SIT for resource monitoring and forecasting over wider areas within limited time-span. The Global Positioning System (GPS), provides three-dimensional positions 24-hour-per-day basis all around the world. This provides latitude, longitude and altitude of any spatial unit. Thus, the GPS has been contributing greatly in the field of spatial mapping. It would be easy to develop predictive modelling capabilities for effective public policies in years to come.

CONCEPTUAL ASPECTS OF SIT

SIT provides an opportunity to interpret various geographical, topographic, socio-economic spatial data in relation to spatial analysis and modelling. The emphasis is on integration of conventional land based and remotely sensed data so as to develop comprehensive data bases suitable for various levels of spatial planning. This can be used for development and planning. In this way, SIT can play an effective role for decision support system for sustainable development and management of the natural resources. The resources are a prime natural capital for human sustenance and development. The development, utilisation and conservation of resources are one of the major priority sectors in the national planning process. Therefore, one can describe the following components of SIT in the context of resource management:

(i) Precise and adequate structured resource inventory or data base,
(ii) Monitoring of space use profile of resource activities etc,
(iii) Analysis of current and time variant resource information,
(iv) Testing of scientific hypothesis about the interaction and functioning of environment and to predict resource potentials, and
(v) Prepare development plan and resource management strategy.

All together three papers were contributed in the section including Jagdish Singh; R. Vijaindra and Afsari Jan; D. Thulasimala and V. Kumaraswami, T. Vasantha Kumaran and R. Radha, S. Sivasankar and A. Udaya Raj.

APPLICATION OF SIT

Geomorphology, Climatology and Water Resources

There is a urgent need for an integrated study of earth systems through space borne, airborne and *in situ* observations with near real time value addition capability for better monitoring and forecasting. A global initiative called GEOSS (Global Earth Observation System of Systems), for which India is a signatory, is pursuing the need to develop global access to a pooled resource of Earth Observation Information. In the coming days, the Indian scientific community needs to focus on the data base content, hazard zonation, vulnerability analysis, speed of information generation, modelling and forecasting possibilities, networking of information, building foolproof interfaces towards information dissemination and efforts towards ensuring that the required information reaches the stakeholders.

Water logging is one of the major land degradation processes that restrict the economic and efficient utilisation of soil and land resources in command areas. Periodic monitoring of command areas helps in analysing the extent of water logging. SIT is a powerful tool, which could be effectively used to study the dynamic behaviour of geomorphic, hydrologic and climatic processes in holistic manner. All together four papers have been included in this volume contributed by S.C. Mukhopadhyay; V.A.V. Raman and S.K. Pal; U.K. De and S.C. Mukhopadhyay; and Chitra Sejwar and S.C. Mukhopadhyay.

LAND USE AND LAND COVER CHANGE

Land use is an important input in the geographical monitoring of rural as well as urban resources of a region. The general land cover classification can be derived from past changes in land use in general and forest cover in particular. Satellite remote sensing can be effectively used for land use mapping and for monitoring forest degradation process in the country. Due to rapid change, the information collected by conventional methods becomes outdated.

Land use suitability analysis helps in taking decision on different land related/based developments. The GIS analysis integrates various natural, anthropogenic and interactive factors to produce three types of maps:

(1) A map showing what land use will cause the least change in environmental processes,
(2) A map showing qualitative predictions of environmental impacts of proposed development.

It also gives certain projects to be carried out and specific environmental actions to be controlled; and

(3) A map showing the best and least suitable locations for those actions.

All together four papers are included in the volume contributed by Afsari Jan and M.R. Vijaindra; Dilip Kumar and Saroj K Pal; Manu Bajpai and A. Velmurugan and T.P. Swarnam.

AGRICULTURE AND LAND USE PLANNING

Land use planning in marginal regions like Himalayan region, Rajasthan desert amongst others requires such mapping. The analysis of land suitability for agriculture in highlands will produce information and maps related to soil erosion, slope, altitude, water availability and nutrient availability etc. The U.S. Department of Agriculture Soil Conservation Service has developed the agricultural Land Evaluation and Site Assessment (LESA) system to determine the quality of land for agricultural uses and to assess agricultural land areas for their economic viability.

The erosional factors prominent in the area, the nutrient quantity and quality of soil have to be taken into account. The cropland use being followed in the area, the irrigation facilities available, the drainage, water logging, salinity and alkalinity status of the area, the vegetation of the area has to be recorded. The mapping has to be done according to degradation status. Interpretation of the results has to be done with a view to suggest various strategies to improve the situation.

Many countries and state governments have attempted to computerise land and other geographical information for municipal operations. Planning departments have made attempts to automate parcel maps. Environmental planners have attempted to automate environmental overlays for land suitability/capability. Such information includes base map, environmental, engineering, plan/profile drainage, parcel/street address tabular data, area tabular data, street network file and area boundary maps etc. The SIT tools allow different data sets in mapped or tabular form to be corrected, interrelated and interpreted in a variety of ways. To make the best use of the SIT, effective networking is essential. It bridges the gap between the users, end-users and data providers. It is important that sharing digitised GIS data as well as socio-economic and statistical data is a key issue to promote spatial technology for sustainable development. All together five papers have been included in the volume contributed by Nandeshwar Sharma, Rais Akhtar, Satyakam Arya and Rubeena.

FOREST AND ENERGY ISSUES

Deforestation for example is a serious ecological problem in the Himalaya. Forest maps of two time periods can provide information of forest cover changes. In many ecologically fragile areas such as mountains, hills and forested regions, fuel wood still continues to be relevant. Various studies reveal that rural firewood consumption was unlikely to be major cause of deforestation because it was largely in the form of twigs and branches. But urban firewood consumption was made up of logs, which are considered a source of pressure on forests. The easing of the supply position of firewood consumption is an interesting development. SIT can be helpful on various aspects of such emerging issues and people's performance on tree planting. Three papers are included in the volume contributed by D. Mukhopadhyay, Praveen Mishra and Rashmi Prakash.

Worldwide experience has shown that SIT is very effective tool for resource management. The new age of microcomputers and increased potential for information exchange aids in its use in India too. SIT offers such capabilities as they integrate multi-sector, multi-level and multi period database. Subsequently, SIT can be used as decision support system identifying and integrating, monitoring and predicting natural hazards, which are the major environment risks in India.

LIST OF CONTRIBUTORS

A. Udaya Raj, Research Scholar, Department of Geography, University of Madras, Chennai.

Afsari Jan, M.Tech. Student, Department of Remote Sensing, Birla Institute of Technology.

C. Chinth Shobana, Department of Geography, University of Madras, Chennai-600 005.

D. Mukhopadhyay, Sparta Institute of Social Studies, A-21, Sector-31, NOIDA.

D. Thulasimala, V. Kumaraswami, Tuberculosis Research Centre, Chennai.

Dilip Kumar, Sr. GIS and Remote Sensing Executive, M.L. Infomap Pvt. Ltd., New Delhi.

(Mrs) Rashmi Prakash, Reader, P.G. Department of Geography, T.M. Bhagalpur University, Bhagalpur-821007.

C. Sejwar, Lecturer, J.S. College, Sikandrabad (U.P.).

Saroj K. Pal, Professor in Physical Geography, Department of Geography, Delhi School of Economics, University of Delhi-110 007.

Satyakam Arya, Reader (Deptt. of Geography) R.R.P.G. College, Amethi Sultanpur, (U.P.) 227405.

M.R. Vijaindra, Lecturer, PG Department of Geography, Bangalore University.

Manu Bajpai, Research Scholar, Department of Geography, Delhi School of Economics.

Nandeshwar Sharma, Prof. and Head, Department of Geography, L.N. Mithila University, Kameshwarnagar, Darbhanga (Bihar).

Praveen Mishra, Carto. Asstt., Department of Geography, Delhi School of Economics, University of Delhi.

Jagdish Singh, Former Prof. and Head, Department of Geog. D.D.U. Gorakhpur University, ex-President NAGI and Director, Grameen Samvikas Sansthan, Gorakhpur.

S.C. Mukhopadhyay, Department of Geography, Calcutta University, Kolkata-700019.

U.K. De, M.Phil Programme in Environmental Science, Jadavpur Univeristy, Kolkata-700032.

R. Radha, Research Scholar, Department of Geography, University of Madras, Chennai.

Rais Akhtar, Department of University of Kashmir, Srinagar-190006 (J&K).

S. Sivasankar, Project Associate, Institute for Ocean Management, Anna University, Chennai.

S.C. Mukhopadhyay, Professor, Department of Geography, Calcutta University, 35, Ballygunge Circular Road, Kolkata-700 091.

T. Vasantha Kumaran, University of Madras, Chennai.

T.A. Rubeena, Department of Geography, University of Madras, Chennai-600 005.

V.A.V Raman, Sr. Lecturer, Department of Geography, Shaheed Bhagat Singh College, University of Delhi, Sheikh Sarai II, New Delhi-110 017.

Velmurugan and T.P. Swarnam, Indian Agricultural Research Institute, New Delhi-100 012.

SECTION I

CONCEPTUAL ASPECTS OF SPATIAL INFORMATION OF TECHNOLOGY

1

SYNCHRONISATION OF VIRTUAL SPACE, SPATIAL SYSTEM AND SPATIAL RE-ORGANISATION FOR RESOURCE MANAGEMENT IN INDIA

JAGDISH SINGH

THE PREMISE

The concept of development has now been enlarged and transmogrified from the neo-classical moorings of economic growth spearheaded by capital formation and accumulation to the process of *enlargement of people's choices* (Sreeten, 1999) and creation of the conditions of *freedom* (Sen, 1999) and thus encompass the *transformation of a society*/community (Stiglitz, 1998). Enhancement of economic efficiency for higher productivity of the factors of production including natural resources, however, remains one of the basic tenets of development. In order to obtain this the origin and spread and assimilation of innovations is a crucial ingredient (Schumpeter, 1934). For the spread and assimilation of innovations in the periphery such as India in the world context and the poverty ridden lagging *BIMARU* states in the national context, is a must; without which the efficiency of the utilization of natural resources either for promoting exports to share one per cent the world trade target or for enhancing the productivity of abundant labour and scarce capital, can not be realized. Free flow of communication through virtual space is a *sina qua non* for the spread and assimilation of innovations originating either from the external or national "cores". As a matter of fact, communication, annihilating physical space, itself constitutes the most potent and cascading series of innovations that are altering the face and pace in every sphere of social life, including the economic.

Also one of the basic pillars of sustainable development is participatory management of natural resources. "This means planning and action is done in a way that facilitates the mobilization and participation of the civil society. Communication is fundamental in development process because it contributes to dialogue and social action. Communication creates opportunities for diverse social actors to come together as equal and participate in the design of proposals. Communication ensures the participation of disadvantaged actors so they influence all aspects of the process—the social, political, economic and cultural. Communication helps to build local capacities, allow people to share their experiences, and participate more effectively. This builds interrelationship between social actors and revitalizes local processes which, in turn, keep the systems that generate local knowledge and alive dynamic communication helps maintain human richness in such processes especially in emergency situations. Communications have always used different media to transmit and receive messages. These are more effective when they encourage an active relationship between actors and mutual responsibilities for the sharing of information that makes dialogues possible. Communication should not be slave to media nor should it be seduced by latest fashion. Communication is not equal to computers, modem and faxes, but to processes of understanding and action". (Hertz and Hoebericht, 2000).

The above statements acquire a special significance in the light of India's march towards *Panchayat Raj* making the *Gram Panchayats* responsible for the formulation and implementation of plans and projects touching core of *Samvikas i.e.* human welfare in terms of survival, security and social dignity of which full employment to every adult in harnessing the natural resources in consonance with the ecological integrity, is the key.

ADVANCE AND SPATIAL COVERAGE OF TELECOM IN INDIA

It is now well known that India, like other developed regions of the world has recently experienced great strides in information technology and thus, entered the arena of virtual space, which serves to annihilate the physical distance separating regions, people and activities. India has, however, failed so far, to effectively harness this revolutionary development to eliminate the glaring spatial disparities in the quality of life of the people. For Indian users internet became a reality in August 1995 when BSNL launched internet services in the metropolises. The BSNL monopoly ended in Nov. 1998 when internet service provider (ISPs) business was thrown open to privy players under the most liberal terms. Even today the license fee for ISPs is just Rs. 1.00 per year while the Cell-phone operators have to shell out lakhs. Fuelled by privatization wave there are today just over 200 million active ISPs covering 340 towns (only) in India. The subscription base was just 3.8 million by Sept. 2002. The user base of net enabled Indians is about 20 million. Cyber cafes and Cyber *Dhabas* number about 12,000 which offer amongst the "cheapest rates" of Rs. 15-20 per hour, in the world.

In Cellular phone service there has been tough competition among the operators. With the slashing of rates by the BSNL and entry of the Reliance in the fray, cell phone service has become almost as cheap as basic phones. Reliance Infocom is poised to offer a complete range of telecom services covering mobile and fixed line telephonic including broad band national and international long distance services, data services and a wide range of value added services through her 60,000 km. long digital and broad band capable terabyte capacity network covering 673 towns and cities in India. The cut-throat competition has led to the slashing of cell-phone rates from Rs. 16.80 minutes, a year before, to Rs. 2.40 with free incoming calls. The commencement of WLL service is now expected to moderate the airtime charges. Reliance Infocom will extend its efforts beyond the traditional value chain to develop and deploy telecom solution for Indian farmers, business, hospitals, Govt. and public sector organisations (Sridhar and Katakam, 2003).

THE SPATIAL CAVEAT

With all this progress and improvement, however, information exchange by the above modern technologies is confined effectively among the cities at the higher rungs. With the introduction of even WLL, it is still very difficult for the rural folk and the inhabitants of the towns to get easily connected to the world outside. As one descends down the urban hierarchy, easy access and availability goes on becoming more and more difficult and exasperating, not to talk of the villages outside the immediate upland of million cities. Thus, the benefits of all the improvements in internet, mobiles phones and the like at progressively cheaper and cheaper rates is reaped by the elite of the city dwellers leaving the vast rural masses including the inhabitants of the towns, as high and dry as ever. For all practical purposes, except perhaps the most urgent exchanges of personal information through PCOs which more often than not fleece the ignorant users. With all the "cheapness" though, it may cost a fortune for the individual user who must spend a substantial portion of his income (per capita net domestic product being < Rs. 10,000 annual in a number of states in '00-0 to use internet, let us say 8 hours per day. This sum goes on escalating as moves down the urban hierarchy because lower down, one has to arrange for private generation of power to get uninterrupted power supply in a situation of prolonged power cuts for duration's increasing with the decreasing size of the settlements. For rural India, therefore, reliable internet and cellular phone service effective enough to make meaningful contribution to the enhancement of productivity is still a pipedream, the rejoicing of the elite internet suffers in the metros, notwithstanding.

"For all the talk of boom in telecom services the tele-density in India is a meagre 4.4 connections per 100 households. The uneven spread across the rural-urban divide is more depressing. Even within the urban setting there are wide disparities in access not only within metropolitan cities but also between the metros and the smaller towns". (Sridhar, 2003). For instance, tele-density in Delhi 13, while in Mumbai it is 11, but in rural India it is just one.

If anything, globalization, liberalisation and privatization spurt has served to accentuate polarization in space, (Banerjee-Guha, 2002) leading to a growing stress on her federal and welfare oriented political set up. This is reflected most apparently in the distortions in the urban-spatial system. The mega and million cities are growing and physically expanding, while the intermediate and small cities and towns are languishing, creating myriads of problems in the space economy. The spatial sprawl of the higher order cities is symptomatic of the contraction in the entire urban system which in turn is an outcome of shrinking of the manpower potential in the national space economy (Singh, 2003). The recent boom in the telecom expansion has so far failed to curb the tendency of polarisation by allowing smooth flow of information and thus help contain the spatial disparity accentuated by the investment of capital in the infrastructurally most favoured locations i.e. the umland of million and mega cities.

BRIDGING THE GAP?

The Govt. opened up the telecom sector to private companies on the premise and stipulation that its action would lay the foundation for greater penetration of telecom facilities and that it would lower cost as well as act as the nuclei for clustering of other infrastructural facilities and social amenities; thus paving the way for urban functions in rural areas as advocated by

Rondinelli (1978), Singh (1979), Sinha (1990) and visualised by the President, Dr, A.P.J. Abdul Kalam is his Vision India 2020.

In 1997, the Govt. introduced a new class of operators using WLL which use radio frequencies to overcome the "last mile" problem by replacing traditionally used "copper cables". It is already evident that private investment is not accelerating access to telecom services. They are interested only in skimming the cream of the Indian market. The regulating authority TRAI has failed so far, to put pressure on them to meet their obligations. Basically, the reach of the Indian telecom network is still synonymous with the reach of the BSNL and MTNL. Since the '90s the cellular operators have established small networks but have ridden the publically owned system that was laid with considerable investment. Reliance has offered STD calls between its own users @ Rs. 0.40 per minute and expressed resolve to make a telephone call as cheap as postcard. Whether this resolve will be translated on the ground of rural India is yet to be seen. Meanwhile, by increasing charges for basic phones with effect from May 1, '03 the Govt. has already tilted the balance in favour of the cellular operators and by design or default, against the rural people.

It is worth recalling that a well balanced and hierarchically healthy log-normal urban system functioning as a well ordered system of "Growth Centres" is *sine qua non* to the kind of resource management conducive to *Samvikas* i.e. holistic development. The spatial organisation, rather re-organisation technology, familiar to geographers: shorn of its positivistic predilections, may be effectively used to synchronise the virtual and physical space provided a *bottom up* rather than *top down* approach is adopted. The information technology revolution has made it so much easier, provided liberalisation and privatisation is properly channellised to expand and effectively cover the entire nation-space, with information network within easy reach of the rural masses. Unless the connectivity revolution serves the cause of the isolated corners of the country, the dream of real connectivity; the connectivity of hearts and minds of one billion Indians can not be realized. The resource creation and management of any kind hinges, thus on the synchronisation of the virtual and real spatial space.

REFERENCES

Banerjee, Guha S., (2002), Critical Geographical Praxis: Globalisation and Socio-spatial Order: *Economic and Political Weekly*, Vol. 37 (44 & 45).

Hertz, &. A, Hoebericht, (2000), The FTPP Potential: A Communication Strategy that makes a Difference; *Forests, Trees and People: New Letter* 40/41.

Rondinelli, D.A. and L.K. Ruddle, (1978), *Urbanisation and Rural Development* Praeger, New York.

Schumpeter, J.A., (1934), *Theory of Economic Development.*

Sen, A., (1999), *Development as Freedom*, Oxford University Press.

Singh, J., (2003), Sustainable Development and Urban System in India; paper presented in the National Workshop on Sustainable Rural and Town Planning (Feb 22-23) Environmental Law Management Capacity Building Project, Ministry of Environment & Forests, Law School, Banaras Hindu University, Varanasi.

Singh J., (1979), *Central Places and Spatial Organisation in a Backward Economy; Gorakhpur Region,* UBBP, Gorakhpur.

Sinha, B.L., (1990), *Urban Functions in Rural Development,* Institute for Rural Eco-development, Gorakhpur.

Sridhar, V., (2003), The Technology Divide, *The Frontline*, Feb. 14.

Sridhar, V. and A. Katakam, (2003), Cover Story, *The Frontline,* Feb. 14.

Stiglitz, J., (1998), Towards a New Paradigm for Development: Strategies, Policies and Processes, 9th Rural Prebisch Lecture, Genf (19.10.98) UNCTAD.

Streeten, P., (1999), Ten Years of Human Development, SPECIAL Contribution, *Human Development Report*, UNDP.

2

CONVENTIONAL TO SPATIAL INFORMATION SYSTEM IN THE EVERCHANGING NATURE OF GEOGRAPHY

M.R. VIJAINDRA AND AFSARI JAN

INTRODUCTION

"Geography can be regarded as a science with the rational development, and testing of theories that explain and predict the spatial distribution and location of various characteristics on the surface of the earth".

The changes in the field of geography can be traced through the following phases, which the subject has gone through to transcend to the current level. The phases are as follows:

	Phases	*Time Periods*
(a)	Pre-historical geographic ideas	5th century BC to 14th century AD
(b)	Period of renaissance and Discoveries	1492 to 1780 century AD
(c)	Contemporary geographic ideas	1780 to 1960 century AD
(d)	Quantitative revolution	1960 onwards
(e)	Modern day trends in geography	1972 onwards

(a) Pre-Historical Geographic Ideas

The history of geographical thought is largely a history of dichotomies and abandoned theories. Throughout history, geographers have developed successive images concerning the arrangement of things on the face of earth. The mental images and explanation of one generation were seldom satisfactory to succeeding generations. There was a continuous search for new and more prefect images and for explanations that conform to contemporary scientific analysis. These changing images and theoretical structures, setup to given them plausibility, make up the content of history of geographical ideas.

In the initial phase, geography was considered as a description of the earth surface and exploration. In the 18th century, it gradually emerged as a discipline of the earth science dealing with man-nature interaction. The contributions made by Hecataeus, Homer, Herodotus,

Eratosthenes and Hipparchus reveal that geography has a longer genealogy than any other science.

During the Greek period, geography was considered to be a branch of knowledge of cosmology, which concerns the study of stars, determination of latitude and longitudes, description of places, mountains, rivers, tribes and nations. This was the period when the habitable world was extremely limited and, therefore, well within the comprehension of scholars. Consequently, the Greek philosophers were well versed in history, philosophy, astronomy, politics, medicine, mathematics, and geography.

The contribution made by the Romans in the various branches of geography are essential to know the process of growth in geography. The Romans with their well disciplined armies; conquered distant countries and they added many of the unknown lands of Europe including Britain to the world map. Polibius and Posidonius contributed to physical geography, while Strabo and Ptolemy were the well – known regional and mathematical geographers respectively. Strabo was a prolific writer and wrote 17 volumes on world geography, while Ptolemy laid foundation of mathematical geography. He developed a concept of 'graticule' which helped in the development of map projection and the science of cartography.

The human process had been like the confused movement of a crowd, whose advance is clearly seen after many stoppages. For about 200 years, there was little contribution of Arabs to geography. Geography and other sciences, however, started with great vigour under the first four of the 'Abbasid Caliphs'. They collected geographical ideas from the Greek and Romans in west and from India and Persia in the east. They added to these the result of their own observations, researchers, explorations and discoveries. They discovered new lands and diffused geography from the Atlantic to the Pacific Ocean. The major contributors were made by Al-Masudi, Al-Biruni, Al-Idrisi, Ibn-Batuta etc., The World Map of Al-Idrisi is shown in *Plate 1.*

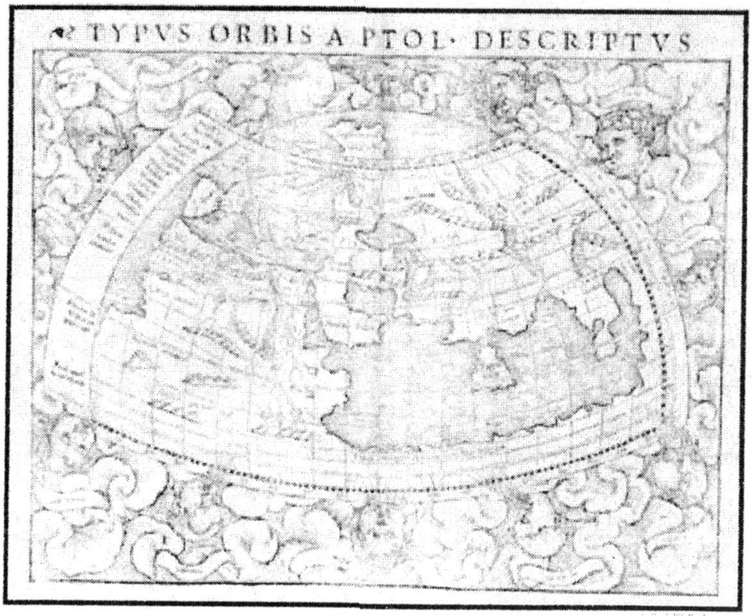

Plate 1: Early World Map Based on Ptolemy

(b) Period of Renaissance and Discoveries

The 'Renaissance' in Europe brought a sudden and complete emancipation from the old modes of thought it provide fresh impetus for widening the existing horizons. There were definite improvements in the art of navigation including wide adoption of magnetic compass. Prince Henry defeated the Arabs and captured the strait of Gibraltar, which made many navigators to explore the West Coast of Africa and the Gulf of Guinea. In 1492, Columbus reached New World (America); Vasco Da Gama reached India in 1498. Here were new developments in cartography – new projections, especially that of Mercator (1569), were invented. In this period most of the erroneous geographical concepts were eliminated.

The shape of the earth was finalized. Varenius and Kant laid the foundation of the classical period. Varenius was the first to distinguish between general and special geography. He imbibed the ideas of Copernicus, Kepler, and Galileo and accepted the heliocentric approach. Kant gave geography a respectable place in the field of learning. *Plate 2* shows the world as known to the scholars in the discovery era.

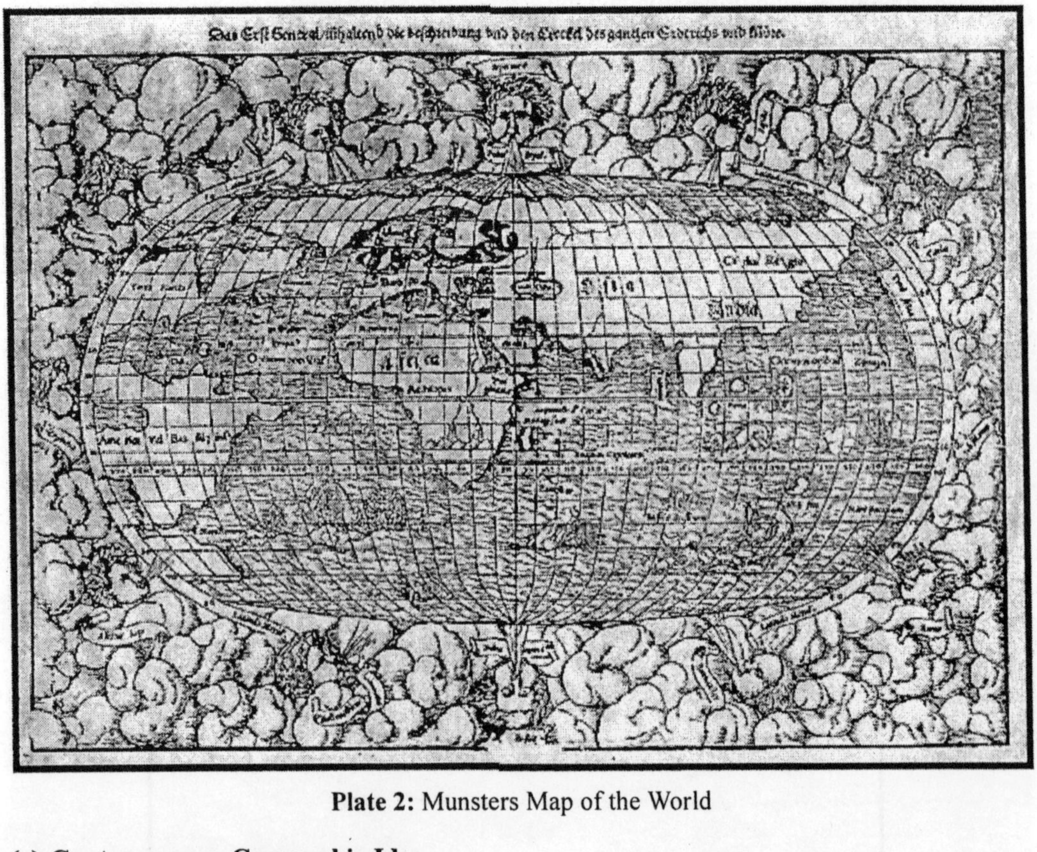

Plate 2: Munsters Map of the World

(c) Contemporary Geographic Idea

After the great age of discovery, two leading German scholars, Alexander Von Humboldt and Carl Ritter made valuable contribution to the fields of basic sciences, humanities and arts.

Based on his famous travels Humboldt wrote his epoch making book 'Kosmos' in which he gave the details of Physical description of earth, landscapes, climate and vegetation. He is considered the founder of plant geography and climatology. Ritter considered geography as that branch of science, which deals with the globe in all its features, phenomena and relations as an independent unit and show the connection of this, unified 'whole' with man. He wrote 19 volumes of geography by name 'Erdkunde' which dealt about science of earth in relation to nature and history.

A major innovation in the world scholarship took place in the 19th century. Geography as a filed of advanced study appeared in Germany in 1874, which was taught by professionally qualified individuals. Within a few decades, geography departments were established not only in Germany, France and Britain but in also other parts of the world. The development of various philosophies and methodologies were adopted by geographers of various schools of thought.

(d) Quantitative Revolution

After the Second World War, geographers, especially those of the developed countries, realized the significance of using mathematical language rather than the language of literature in the study of geography. Consequently, empirical descriptive geography was discarded and greater stress was laid on the formation of abstract models. Geography now began to provide accurate, orderly, and rational descriptions and interpretation of the variable character of the earth surface. *Plate* 3 shows the world in the second half of the nineteenth century.

Plate 3: Al Idrisi's World Map

After this revolution, quantitative techniques and general system theory have been used quite extensively in geography. Statistical methods include the theory of frequency distribution, Bayesian theory of hypothesis testing. Linear model, entropy-maximizing modelling, graph theory etc.

(e) Modern Trends in Geography

With the advent of quantitative revolution, the modern science and advanced technology was also bringing in new innovations in the field of geography. Now geography is also being rechristened as SPATIAL SCIENCE. The collection of data without enduring physical hardships became more common. The main new techniques, which are aiding geography in becoming a more precise science, are as follows:

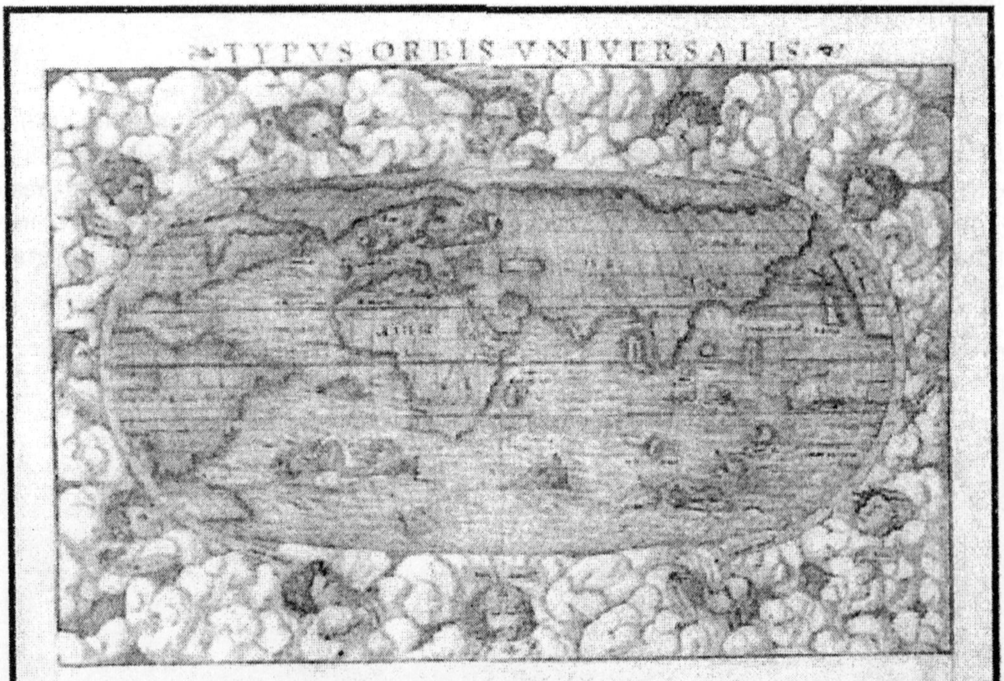

Plate 4: The Discovery Era World

(f) Photogrammetry and Remote Sensing

The art, science and technology of taking reliable measurements on photographs object and environment is known as photogrammetry. Aerial photography came in to practice when Louis Daguerre announced his direct photographic process in 1839. The invention of aeroplane by Wright Brother in 1902 provided great impetus to the emergence of the modern aerial photogrammetry. Since then, the science of photogrammetry has developed in to instrumentation, cameras and films.

Plate 5: Satellite Image of the World

Remote sensing is the science and art of obtaining information about an object, area or phenomenon through the analysis of data acquired by device that is not in contact with the object, area, or phenomenon under investigation. Remote sensing of earth resources encompasses all information on or below the Earth, collected from a distance, such as aerial photographs and satellite images. Photographic interpretation involves the study of photographic images, while remote sensing, which is relatively a new technology, includes not only the analysis of photographs but also the use of data gathered from a wide range of remote sensing sensors and scanners on board the satellite. Remote sensing images, obtained from the orbiting satellites, are capable of providing quantitative as well as qualitative information about objects. Remote sensing activities started in the 1950's. The usefulness of image for mapping and other application increased in 1960, with the launching of the first US meteorological satellite TIROS-1. The launching of the first Earth Resources Technology Satellite (ERTS) by USA in July 1972, marked the beginning of space development activities, including the use of remote sensing images for various natural resources applications. Aerial photo-interpretation and remote sensing both have been employed in natural resource mapping, management and planning. Today, remote sensing affords a practical means for frequent and accurate monitoring of the Earth's resources, literally on a global basis. *Plate 5* shows the satellite image of the world of today.

DIGITAL INFORMATION SYSTEM AND DIGITAL CARTOGRAPHY

It is primarily advances in computer technology that have enabled the development of GIS. The decreasing cost of computer power over the last two decades has been one stimulus for wider use of GIS. The initial developments were technical and were aimed at developing a set of spatial data handling and analysis tools that could be used with geographical databases for repeated problem solving. The technical issues for early consideration included digitizing, the development of data structures to allow updating and analysis. Main development of 1960's was the layer approach to map data handing in CGIS, and pioneering work by the US Bureau of Census that led to the digital inputs of the 1970 census. The development in GIS from the ESRI is mentioned in the given Table.

Table

Years	*Development of ESRI*
1969	ESRI founded by Jack and Laura Dangermond as a privately held consulting group in Redlands, California
1970's	ESRI involved in application such as site selection and urban planning which improved technical and applied aspects of GIS
1981	ARC/INFO GIS for microcomputers launched. Later the product was shifted to UNIX workstations and PC's
1986	PC ARC/INFO was launched
1991	ARC/VIEW was launched
1992	ARC/CAD launched
1996	ARC/INFO for windows NT was launched

Source: Adopted from ESRI, 1996.

The stage in which the progress in GIS can be traced is as follows:

Stage and Date	*Description*	*Characteristics*
Early 1960's 1975	Pioneering	• Individual personalities important • Mainframe-based systems dominant
1973-early 1980's	Experiment and Practice	• Local experimentation and action • GIS fostered by national agencies • Much duplication of effort
1982-late 1980's	Commercial Dominance	• Increasing range of vendors • Workstation and PC systems becoming available • Emergence of GIS consultancies and bureaus • GIS World trade journal launched
1990's	User dominance, vendor competition	• Embryonic standardization • Systems available for all hardware platforms • Increasing use of PC and networked systems

Sources: Adopted from Tomlinson, 1990, Coppock and Rhind, 1991.

Digital cartography is a method of storing and editing of maps using a computer. The development of computerized GIS is closely linked with computer cartography. SYMAP (Synagraphic mapping) was probably the first commercially available automated cartography package. It was first marketed in the USA in about 1965 and could produce crude but useful choropleth maps. Production of map was quick but the quality of output was not good but now it is not just good but excellent.

ANALYSIS

"When a coin tossed high in the air, it is quite impossible to tell on which side it will land on ground. Just in the same way it is impossible to tell whether the conventional or modern geography is superior. In the case of a coin when it hits the ground its fate can be told but, its quite impossible for a subject as the information coupled with new technology is flowing in for infinity".

Though it is highly complex a simple analysis has been attempted to know its positive and negative aspect and it has been presented in the table given here:

Conventional Geography	*Modern Geography*
The map from the pre-technical era gives information in the form of signs and symbols, which are difficult to decipher for a common man.	The aerial photography and satellite image are the direct view of ground reality which is very clear and exactly what one finds on the earth surface which can be deciphered easily.
The hardships encountered in this era for collecting ground reality was depending on the accuracy of the surveyor or traveller with no proof of authentication except cross check from another traveller.	The satellite image provides synoptic view, which is clear to the eye. It is acquired by the sensors and cameras mounted on air and space crafts. There is not physical hardship in this and the data befits the global standards and is authenticated by space organization.
The preparation of maps took year in the making and by the time they were finished there would be many changes in ground conditions and the map which finally came to light had only a limited area of influence.	The maps can be prepared very fast with the help of satellite images in a short time and the monitoring of earth to observe on a regular basis can be done as the satellites have temporal resolution.
The map when created with all the information put in to it is a hard copy, which can neither be rectified nor updated.	The digital maps created in GIS can be corrected, updated to suite the users need and specification, which was unthinkable few decades ago.
There was a flair and dedication towards appreciating the environment and to describe them in great detail in a very flowery way.	The current science is highly quantitative and non-descriptive. Today with the number of scientific research going on no one has time to think about the surrounding least write about it.
The access to the knowledge about inventions and discoveries was highly restricted to the scholarly circle and a few political bigwigs.	The access to the knowledge about science and technology now reaches each and every nook and corner of the globe with mass media to each and every human being that has the zest for knowledge.

One outstanding sad aspect to this whole process of evolution of geography is that from past to present the new findings have always been used for war and not for peace. My guess to where all this technology and modern science is going to lead mankind is as good as any one else's.

CONCLUSION

A geographer was believed to be jack of all but master of none. It is certainly time for the masses to change their opinion, because geographer is the one and only master of spatial science.

REFERENCES

1. Agarwal and Garg, '*Text book on Remote Sensing in Natural Resources Monitoring and Management*', Wheeler Publishing, 2000.
2. Heywood *et. al.*, '*An Introduction to Geographical Information Systems*', 2000.
3. Husain Majid, '*Evolution of Geographical Thought*', Rawat Publication, Third edition, 1999.
4. WEBSITES
   (a) www. Gisdevelopment.com
   (b) www.esri.nationalgeographic.com
   (c) www. Africanhistory.com

3

GIS APPLICATION IN MEDICAL GEOGRAPHICAL ANALYSIS

D. THULASIMALA, V. KUMARASWAMI AND T. VASANTHA KUMARAN

INTRODUCTION

Geographical Information Systems (GIS) are defined as computer based systems for entering, storing, analysing and displaying digital geo-referenced data sets.

GIS is a powerful tool for geographical analysis for almost any health discipline. GIS technology is one of the fastest growing technologies in health today, helping professionals in the complex health industry, in managing their information to make better decisions. At present, GIS is seen primarily as a research tool in the field of vector-borne disease; however, analysis procedures continue to develop at a rapid pace. However, the use of GIS in decision support has also become an area of growing interest. GIS is defined as a constellation of software and hardware tools, capable of integrating digital images for the purpose of dealing with geographically localised data.

Application of remote sensing images to individual development projects or programme raises further considerations of scale. High-resolution images are useful in mapping the project area and determining potential hazards (Walter, 1994). Low-resolution images are of limited value over a small area.

GIS AND MALARIA

Malaria is still recognised as one of the most prevalent and debilitating of the tropical diseases, with nearly 2.1 million people at risk of infection (WHO, 1993). The WHO/GLOBAL strategy identified the strengthening of local capacity in basic and applied research of a country's malaria situation, in particular the ecological, social and economic determinants of the disease. In ancient India, malaria was known as the 'king of diseases'. Malaria and famine formed the vicious cycle of sickness, death and poverty, and thus malaria was largely responsible for the poverty of nations.

Measures to improve health and quality of life are gaining greater attention, together with the need to protect and improve the environment. One of the most important questions asked in public and environmental health today concerns the type of instruments that can be used to devise quick, reliable and scientifically valid methods for rapid assessment of the situation in order to assist in health research and planning, as well as monitoring and evaluation of health programmes. GIS can be defined as an automated information system based on the methodology of geographical database management and query. Chennai City, with its high incidence of malaria and unabated growth of its urbanising areas, has given rise to urgent need for newer methodologies to assess the situation. Malaria being a local phenomenon, the intensity of its transmission, prevalence and distribution of parasites are determined by the local *malariogenic* conditions, which differ from neighbourhood to neighbourhood and even from divisions to division in the same locality. Moreover, the developmental activities in recent years have brought in major morphological changes in the erstwhile-established ecosystems. Therefore, it is essential to assess the situation for the spatial distribution of malaria based on the *malariogenic* factors responsible for its resurgence or increase, and also decline in its incidences, over time.

This calls for a new approach and the approach for understanding the spatial patterns of the disease and gaining insight for developing a *'spatial strategy'* for urban malaria control and prevention. Many GIS systems have the capability of incorporating aerial photographs, satellite and radar imagery into their data layers. The process is simple, as the imagery is scanned or read off a data type. GIS is a tool, which allows efficient synergism of spatial data in the form of maps and tabular data.

The integration of information derived from remove sensing techniques with other data sets, both in spatial and non-spatial formats, provides tremendous potential for characterization and analysis of natural resources a swell. GIS technology offers efficient utilities for handling spatial data. It allows for both the spatial and non-spatial data base organisation, mainly as attributes of the spatial elements, multi-layer integration and modelling of different parameters to arrive at inputs to decision-making and visualisation of information in newer perspectives. The greatest advantage of spatial representation is the perception of contiguity and connectivity that the data convey, which are an important element for planning and management (Raj *et. al.*, 1997).

Researchers have been developing automated tools for efficient storage, analysis and presentation of geographical data (Aronoff, 1989). The GIS systems, resulting from the demand for data and formation of a spatial nature, are widely used across varied scientific fields. GIS offers the ability to process quantities of data beyond the capacities of manual systems. Data are stored in a structured digital format, which permits rapid retrieval and use. In addition, data are quickly compiled into documents, using techniques such as automatic mapping and direct report printouts (Bernhardsen, 1992).

GIS facilitates the integration of quantitative malaria detection and control data with data obtained from maps, satellite images, and aerial photographs. Frequently socio-economic data and qualitative information on health facilities have a spatial basis (for example, Tulasimala and Sivagnanam, 1998) and can also be integrated with malaria data from the same areas. The integration of operational and logistical data for malaria control programme planning with epidemiological data serves to strengthen both the epidemiological analysis and the planning and execution of control programmes.

Malaria stratification aids in the development of community-based malaria control programme by accumulating past experiences with an solutions to different factors associated with malaria outbreaks (Orlof, 1986). Stratification can also point to the existing inequalities in resources, allowing for more equal and homogeneous distribution of available resources (Kadt and Tasca, 1993).

The ability of GIS to deal with large data sets and to incorporate satellite images increases the feasibility of studying the environmental determinants of malaria. The epidemiological mapping of high-risk areas of malaria transmission has helped national level malaria control programme to recognise those populations and geographical areas, where it is possible to identify the main determinants of malaria morbidity and mortality.

One of the main functions of GIS is the selection of areas according to some specified characteristics, in a search for regularities. These characteristics may be spatial or non-spatial, from a map or data base within the GIS. Data may be selected according to geographical criteria or content. Spatial queries may be built to stratify areas according to inundation potential, proximity to treatment sites, availability of transport, incidence of malaria, quality of housing and so on. The stratification of areas according to their epidemiological situation is also useful to developmental agencies in implementing health projects.

Staff members of malaria control programme are familiar with maps, using them for their daily activities. Maps can consequently be used to communicate ideas and explanations about the determinants of malaria and strategies for control. A map's potential is exploited as the medium of communication between a control programme and the public.

Statistical operations are performed primarily on attribute data, but may also be applied to some types of geometrical data. Most GIS supports a range of statistical operations. Pattern recognition and statistical modelling are also incorporated in some GIS (Bernhardsen, 1992), allowing for the identification of spatial and/or temporal clusters of malaria occurrences. The relation between these clusters and environmental or socio-economic characteristics can then be investigated.

GLOBAL CHANGES AND MALARIA

Changes in the global environment will lead to changes in the pattern of malaria occurrence. These changes are currently being studied and modelled, and their impact on malaria can now be examined. The occurrence of malaria under different land occupation susceptible can be used as guidelines for new development projects in areas susceptible to malaria.

The WHO Technical Report No. 839 (1993) recognizes that a change from specialized information systems for malaria towards integrated health information system is essential, if the general health services are to take full responsibility for managing malaria (WHO-TDR, 1993). An integral health information system will allow the malaria to be related to other health problems.

The digital image is fed into the computer, where it is stored. The digital images can then be displayed and further processed to extract the desired information. They may also be integrated with other types of data and information within a GIS. By using the data from the different wave bands, it is possible to identify and track environmental characteristics and changes useful to the study of malaria and other tropical diseases.

THE OBJECTIVES

As this study is only a small part of a larger on-going study on urbanisation and vector control research, the objectives given below are only partial. They are:

(1) To develop a GIS for malaria so as to enable us to identify areas of high malarial incidence, towards developing a spatial strategy for control and eradication; and
(2) To examine the trend in incidence over the years so as to understand the reasons for resurgence.

For mapping, however, the data used is only for one year 2000, although the pattern of change in malarial incidences is discussed in detail.

METHODOLOGY

The collected information is processed and analysed using the GIS package MAPINFO. The spatial database and the attribute database for the entire Chennai city have been developed using Microsoft EXCEL and adapted to MPAINFO as well as SYSTAT. By using MAPINFO, four maps of Chennai city have been prepared as follows: *Population* by city wards; distribution of *Plasmodium Vivax* by city wards; *Total positive* cases by city wards; and *Annual Parasite Index* by city wards. The data mapped, primarily for demonstrative purposes, relate to the year 2000, as the one-year data have been made available by the Corporation of Madras for the purpose.

URBAN MALARIA: A CASE OF CHENNAI: METROPOLIS

Chennai is the fourth largest city in India. For administrative purposes, it is divided into 10 zones and 155 administrative wards. Chennai is spread over 174 km^2. It has an average slope of 150 cm to 240 cm towards east-southeast. It is a lowlying terrain. The climate of Chennai is tropical maritime monsoon type. The mean minimum temperature rarely falls below 20°C, while the maximum temperature seldom crosses 37°C. The city is found sprawling towards the inland and the south with a higher growth rate of population.

The total population of Chennai city is 7.2 million and it is spread over an area of 174 km^2. In Chennai city, the population grew at 27.35 per cent during the decade 1971-1981 and at 15.82 per cent during 1981-91. It is even lower a growth rate in the last decade. But migration into the city is adding millions over decades.

Figure 1 shows the distribution of population. The very high concentration of population is found only in five wards, namely, Vyasarpadi (S), Perambur (N), Nagammaamaiyar Nagar (S), Dr. Sathyavani Muthu Nagar and Sathyamurthy Nagar. The highest ward population is above 40,000 and it is observed in five wards as well. The lowest concentration is population is in the Northwestern, Southern and Central parts of the city and this is less than 20,000 in each of the wards. Two of the northern border wards are also seen with low population. The distribution of less than 20,000 is in eight wards and they are 1, 2, 85, 93, 95, 101, 102, and 110. Varying wards have different levels of concentrations. For example, 20,000 to 25,000 populations are found in twenty two wards and 25,000 to 30,000 are in sixty one wards. The moderate concentration (30-35,000) and highest concentration (35-40,000) of population are found in seventy-four wards.

MALARIA: THE DISEASE, ITS SPATIAL PATTERNS AND VECTORS

Malaria is a communicable disease caused by sporozoan parasites of genus plasmodium and transmitted to humans by certain species of infected, female *Anopheline stephensi* mosquitoes. The disease is characterised clinically by three stages (Gubler, 1998, for example).

> The first is the **Cold Stage**: Sudden onset of fever with rigour and sensation of extreme cold accompanied by shivering;
> The second is the **Hot Stage**: Patient feels burning hot and casts off clothes covering him. There is intense headache; and
> The third stage is the **Sweating Stage**: Fever comes down with profuse sweating. There is enlargement of spleen and secondary anemia with a tendency to relapse.

Malaria in India was essentially a rural problem. During the launching of the National Malaria Control Programme (NMCP) in 1958, and the National Malaria Eradication Programme (NMEP) in 1958, no proper attention was given to tackling the urban malaria problem. Malaria in urban areas showed only a gradual increase. In Tamil Nadu, more than 90 per cent of the total malarial incidence was recorded from urban areas during 1963-67 when the malaria eradication was at its peak in the city. On an average, about 70 per cent of positive cases of malaria was recorded from the urban areas of Tamil Nadu between 1961 and 1977.

Of the 10 urban areas under the Urban Malaria Scheme (UMS) in the state of Tamil Nadu, the problem of malaria in Chennai is acute. It was observed that 56-77 per cent of the total malarial incidence between 1980 and 1996 in Tamil Nadu was contributed by the urban areas alone. The malarial incidence from the state of Tamil Nadu between 1977 and 1966 ranged between 35 to 75 per cent of the total incidence in the state. The city was brought under UMS during 1971-72.

TRENDS IN THE INCIDENCE OF MALARIA IN CHENNAI 1979-2000

The number of malaria cases recorded in the city was 617 in 1973 and 2,560 in 1974. During 1985, however, there was a sudden spurt in the incidence with as many as 36,207 positive cases were recorded, demonstrating a 15 – fold increase over the previous year 1974. During 1975-84, the number of positive malarial cases fluctuated between 30,000 and 45,000. As many as 57,376 positive cases were recorded during 1985. Thereafter, the number of cases declined to 31,126 in 1986 and then increased to 45,622 cases in 1989, reaching its peak at 76,749 total positive cases of malaria during 1993. A total of 45,930 cases were recorded during 1996. The percentage of *Plasmodium falciparum* was below 8.0 per cent, for all the years except during 1991, when it was 12.0 per cent with a total of 8,024 cases and during 1992 when it was 10.9 per cent with a total of 7,858 cases. The contribution of *Plasmodium falciparum* cases from Chennai city in respect of the state total was also observed to be between 54 per cent and 75 per cent during 1980-96 (see Hyma and Ramesh, 1980).

Data on total cases recorded during 1979-2000 showed that the incidence both increased and decreased over the years. In the early 1980s, the incidence increased by a high of 24 per cent in 1981 over the year 1980. The year 1983 showed a marginal decline of 0.36 per cent only, to progressively increase in the next two years to 8.0 per cent in 1984 and 6.0 per cent in 1985 over the previous years. For the year 1986, the decline in total cases was drastic by 24.0 per cent. In the next few years, in the order of decline, it was 33 per cent in 1989, 31.0 per cent in 1991, 12.0 per cent in 1990, 11.0 per cent in 1988, 8.0 per cent in 1992 and 6.0 per cent in 1993.

The expectation is that the total cases recorded will fluctuate but malaria cannot be eradicated or controlled unless certain social and spatial strategies are adopted stringently and with telling effect. With the current trends and behaviour, resurgence has greater possibilities. This calls for efforts on a war footing, both spatially and socially, over the next few years.

Records have that the *Plasmodium falciparum*, the fatal vector, has also been raising its ugly head with greater increases over some years than declines in other years. *Plasmodium falciparum* was on the increase for a period of 6 years during 1979-84, touching a high incidence of 3,558 in 1984 but the rate of change was 339 per cent in 1981 and 101 per cent in 1982 over the previous years, respectively. Between 1985 and 1991, the incidence of *Plasmodium falciparum* registered increase, reaching 105 per cent in 1991. There was again a period of decline for 4 constructive years and then a period of fluctuation for the next 5 years. *Plasmodium falciparum* has the potential for resurgence as well and it will be a dreaded vector in the years to come, as it is fatal.

Total Positive Cases

From the analysis, the very highest number of positive cases of malaria are found in three wards, namely, Korukkupet, Sowkarpet and George Town (Fig. 2). The lowest number of positive cases of malarial incidence is found in 110 of the 155 wards. They primarily fall under less than100 cases category. The range of 100-500 cases is found in thirty eight wards and that of 500-1000 cases are found in four wards, namely, M.G.R. Nagar, Navalar Nedunchezhian, Nagar, Vallal Seethakathi Nagar, and Dr. Radha Krishnan Nagar (S).

Plasmodium Vivax

Malaria has seasonal outbreaks, in most parts of India, even though it occurs throughout the year with varying intensities. The maximum prevalence of *Plasmodium vivax* cases is found in three wards, namely, Korukkupet, Sowkarpet and George Town (Fig. 3). The lowest incidence of positive cases is found in one hundred and twelve wards (112). The contiguous stretch running Northwest-Southwest of Chennai city falls under the less than 100 cases category whereas 100-500 cases are found in 36 wards. The range of 500 to 1,000 *positive vivax* cases has been observed at Navalar Nedunchezhian Nagar, MG.R. Nagar, Vallai Seethakathi Nagar and Dr. Radhakrishnan Nagar (S).

ANNUAL PARASITE INDEX (API)

Figure 4 shows the Annual Parasite Index for the year 2000. There wards are found to have very high disease prone (> 5) areas: Korukkupet, Sowkarpet and George Town. The striking point is that the three wards are repeating in all mapped patterns again and gain. These areas are the high risk areas. The lowest incidence of (<1 – 2) positive index is found in ward numbers 29 and 31. The range of Annual Parasite Index 2 – 3 is found in thirty wards. Only seven wards fall under the 4-5 category.

The occurrence of malaria in Chennai appears to be long standing. As early as 1912-15, larvivorous fishes were introduced in the fresh water wells of George Town in the city as this happened to be a notorious area for malaria. The transmission of malaria in Chennai is perennial and the peak transmission periods are between July-August and October-November.

The rapid rate of urban development, particularly sub-urbanisation in the periurban areas, has attracted migration to the city and aggregation of labour force from various parts of India, besides new slums and expansion of the older areas. Reports of malaria epidemic resulting from the migratory population are not uncommon. Moreover, the reports on the development of resistance to chloroquine in *P. falciparum* in Chennai city are a cause for concern.

MALARIA CONTROL AND PREVENTION

Malarial incidence may be reduced in the next few years in most of the urban neighbourhoods, if only environmental improvements in the city of Chennai are carefully made by reducing more intensive irrigation in rural and urban fringe areas and abundance of water use in the inner urban areas. In urban areas, if the incidence of malaria could be brought down without insecticides and drugs in major part of the country, it would certainly be a great step forward. Regions, which have no problems, should be taken up for intense research, even if the methods of control and prevention to be used in them are expensive for the country as a whole. Inputs needed are the technical know-how; finance, and manpower.

While technical know-how for eradication and control of malaria is available, finance has been a major problem. Most municipalities find it difficult to address the financial problem because of low resources. International efforts are such that finances may not be difficult in the long run. Chennai city has found money from the World Bank for its programme of UMC. The most important requirement is manpower, to execute control measures planned for the city by the Corporation. The shortage of manpower in the Corporation, for example, to attend to the larvicidal fish programme and mosquito bed nets among the public is proving to be the undoing. Whatever has been done all these years has been ineffective, requiring new thinking and concerted effort.

India is a signatory to the Global Malaria Control Strategy signed in Amsterdam in October 1992. In keeping with this international commitment, the National Malaria Control Strategy (NMCS) has been drafted to supercede the modified plan of operation. To control the vectors, it should include all types of measures, permanent and recurrent, necessary to destroy the vectors and their breeding in the surroundings of urban as well as rural areas. Environmental improvement should necessarily be implemented in urban areas, especially in slum and non-slum areas alike. What was lacking till now has been the will to undertake it and a certain amount of spoon-feeding of the local administration by the Central and State Governments, which had taken away the initiative from the local bodies.

To arrest the rising trend, Urban Malaria Scheme (UMS) was launched in 1971-72 in 131 urban areas in 18 States and Union Territories. Severe steps have to be taken as has been done in respect of malaria in Mumbai. Here it was under effective control but in the 1990s, malaria has surfaced as a major public health issue. New Mumbai has remained by far endemic as various nodes undergoing intense construction activities supported the breeding of *Anopheline Stephensi*.

In Tamil Nadu, Chennai contributes up to 60 per cent or more malaria cases to the total cases in the state. The Tamil Nadu Government promotes control of the vectors; they are using anti-larval methods, often supplement by thermal malathion fogging. Fogging is expensive and unproductive as a routine method of malaria control, but it is in great demand because of its visibility and social acceptability. To control mosquitoes rather than emphasising species control and sanitation, that is, the control of the vector. *Anopheline Stephensi*, field operations should be directed to control mosquitoes.

CONCLUSION

The paper has demonstrated the usefulness of GIS in mapping malaria, in terms of distribution of incidences of positive cases, *plasmodium vivax*, and API. Having demonstrated the use of GIS in malaria research, it is possible to extend the use of the tool and technology in other contexts as well, especially vector-borne diseases such as filariasis, dengue and Japanese encephalitis. The mapping accomplished can be used further in devising spatial strategies for controlling and preventing the diseases.

REFERENCES

Aronoff, S. (1989), *Geographical Information System: Management Perspective*, WDL Publications, Canada.

Bernhardsen, T. (1992), *Geographical Information Systems*, Viakt IT, Arendal, Norway.

FAO (Food and Agriculture Organisation) (1990), *Satellite Remote Sensing in Support of Early Warning and Food Information Systems—Strengthening National Early Warning and Food Information Systems in Africa*, Annex 7, FAO Workshop, Accra, Ghana, 23-26, October, 1989.

Gubler, Duane, J. (1998), Resurgent Vector-Borne disease as a Global Health Problem, *Emerging Infectious Diseases*, 4 (3), 442-450.

Hastings, D.A. and D.M. Clark (1991), 'GIS in Africa: Problems, Challenges and Opportunities for Co-operation', *International Journal of Geographical Information Systems*, 5, 29-39.

Hyma B. and A. Ramesh (1980), 'The Reappearance of Malaria in Sathanur Reservoir and Environs: Tamil Nadu, India', *Social Science and Medicine* 140 (4), 337-344.

Kadt, E. and Tasca, R. 1993. Promovendo a equidate: *un novo enfoque com base no setor saude. Editora Hucitec e cooperacao Intaliana em Saude*. Sao Paulo-Salvador, Brazil.

Orlof, V.S. *et. al.*, (1986), 'The Concept of Stratification of Territories and its Practical Implications,' *World Health Organization*, Geneva, Switzerland, WHO/MAL/86.1032.

Perry, B.D. and R.L. Kruskas, (1993), 'Information Management Technology to Strengthen Veterinary Services: An Appraisal of the Systems Available and their uses in Developing Countries', Expert consultation on the need for information systems to strengthen veterinary services in developing countries, Report, FAO, Rome, Italy.

Rajan, M.S. (1991), 'Remote Sensing and Geographical Information Systems for Natural Resources Management', *Asian Development Bank*, Manila Philippines. ADB Environment Paper No. 9.

Thulasimala D. and N. Sivagananam, (1998), 'Health Hazard in Northern Tamil Nadu—Analysis for Estimating Health by Social Determinants', *Region, Health and Health Care* 3 (2), 11-22.

Raju, U., K.S. Ramesh, B.K. Ranganath and S. Adiga 1997: Role of satellite remote sensing and geographic information system in establishing APIB, Proceedings of the one-day Workshop on *Agro-climatic Planning and Information Bank*, Regional Remote Sensing Service Centre, ISRO, Bangalore: 8-11.

Walter, L. 1994: Natural hazard assessment and mitigation from space: the potential of remote sensing to meet operation requirements, in G. Wadge (ed): *Natural Hazards and Remote Sensing, Proceeding of Conference* held on 8-9 March at the Royal Society, London, UK: 7-12.

WHO 1993: *Report of a WHO study group on the Implementation of the Global Plan of Action for Malaria Control, 1993-2000.*

WHO Report Series 839, Geneva, WHO Indian National Science Academy (1979). *Symposium on Insect Vector Biology* 739-741.

WHO-TDR 1993: TDR News. Newsletter No. 43. WHO Geneva, Switzerland.

4

INFORMATION TECHNOLOGY FOR TOURISM RESOURCE MANAGEMENT: APPLICATIONS OF GEOGRAPHIC INFORMATION SYSTEMS (GIS)—AN OVERVIEW

R. Radha, S. Sivasankar and A. Udaya Raj

INTRODUCTION

Tourism is largely a natural resource based industry and, as such, it affects air, land and water. It can damage natural systems if its planning, development and operation are not properly managed. Tourism that is properly planned and managed, however, can minimise the impact on the environment – and even benefit the environment. Ecotourism, which is defined as 'Responsible travel to natural areas that conserves the environment and sustains the well-being of local people' (TIES) has developed and grown very rapidly. Ecotourism is starting to redress the imbalance of conservation and development by helping the rural communities conserve their resources and benefit economically. However, little effort has been directed towards environmental issues. To minimise the environmental impacts of tourism, particularly the diffuse impacts in conservation reserves, one needs to understand the system thoroughly, as a large volume of information requires consideration. Therefore, quantitative measurements of the environmental impacts of tourism in conservation reserves and fragile environments are critical for resource managers and planners.

As mentioned above, information is the lifeblood of the tourism industry and it is inevitably influenced by the technological revolution. Therefore, the effective use of latest technology such as, Information Technology (IT) is crucial to its competitiveness and prosperity. Timely and accurate information, relevant to the consumer's needs, is increasingly the key to satisfaction of tourist demand and the ability of tourism enterprises to offer services that meet the customer's demand. IT propels tourism demand and supply by providing effective tools both to the consumers and suppliers. Geographic Information Systems (GIS) is one such tool of Information Technology that facilities the identification, development, management and distribution of resources. It therefore becomes an essential partner, which increasingly determines the interface between consumers and suppliers globally.

This paper attempts to describe the applications of IT and GIS from tourism management point of view. As this work gives only a theoretical overview of the areas of application, this has been arranged in the following order. The first section deals with the tourism resources; second section explains the varied applications of IT, in specific, GIS; and the last section elaborates the ways in which IT and GIS can be used for tourism sector.

TOURISM RESOURCE MANAGEMENT

Research forms the backbone of any scientific intervention—tourism management is no exception. In the 1980s, alternative forms of tourism began attracting the interest of governments, communities and scholars alike. These were given a raft of names—"nature tourism", "soft tourism", "responsible tourism", "green tourism", "ecotourism" (Schaller, 1999), but all were seen as alternatives to mass tourism. Among these various labels, the term "ecotourism" has become prominent, although a consistent definition is by no means found, even among scholars (Schaller, 1999). Most definitions however, incorporate concepts associated with sustainable development. Sustainable tourism development is based on the concept of Ecologically Sustainable Development (ESD), which means using, conserving and enhancing the community's resources so that ecological processes, on which life depends, are maintained and quality of life for both present and future generations is increased.

Many scholars now agree that ecotourism should require a two-way link between tourism and environmental conservation (Valentine 1993; Cater, 1994). As the understanding of the close relationships between tourism and environmental conservation increases, researchers are calling on ecotourism industry to incorporate the technological development so as to promote tourism as a much more organized economic activity.

Ankomah and Crompton (1990) identified five factors inhibiting the development efforts as negative market image, lack of foreign exchange for capital development, lack of trained personnel for tourism, weak institutional frameworks for planning and management and political instability. Sournia (1996) suggests that the reasons for the lower levels of resource utilization include less visible resource concentrations, weak national transportation networks, inefficient hotel facilities, poorly trained tourism staff, weak marketing and a lack of tourism infrastructure in the parks.

There are other factors with complex inter-relationships that demand for the usage of sophisticated systems for better management of the resources. Some of these are: A Natural resources, Cultural resources, Tourist Facilities, Accessibility, Tourist demand. Infrastructure Environmental quality, Local community, and Historical Occupation among others.

THE TECHNOLOGIES: IT AND GIS

IT incorporates the software, hardware, information, management and telecom systems to enable the processing of large volume of data and also to maintain the flow of information within and between organizations. Therefore IT should be regarded as a synthesis of electronics, computing and telecommunication technologies (Poon, 1993).

GIS provides a tool for effective and efficient storage and manipulation of remotely sensed data and other spatial and non-spatial data types for both scientific management and policy-oriented information. This can be used to facilitate measurement, mapping, monitoring and modelling of variety of data types related to natural phenomenon. GIS an improve the quality and power of analysis of natural resource assessment, guide development activities and assist planners in the selection of measures. This section gives a brief introduction about the Information Technology with an understanding that the readers are already aware of the same.

Geographic information systems are becoming routine analysis and display tools for spatial data, used extensively in applications such as land-use-mapping (for urban planning purposes), transportation mapping and analysis (for determining efficient transportation routes for deliveries and emergency response), demographic analysis (for facilities location), utilities and infrastructure mapping and multiple applications in natural resource assessment (including water quality assessment and wildlife habitat studies).

GIS allows efficient and flexible storage, display, and exchange of certain kinds of spatial data, as well as potential interface opportunities for a variety of quantitative spatial analysis models. Users include: federal, state and local governments and their agencies, private firms, non-profit organizations, grassroots and community groups, universities, and research institutes. The following are the process involved in the application of GIS for any kind management.

1. **Documentation.** This process feeds GIS system with graphical and attributive information.
   - *Recollection,* by in-site registration, surveys and institutional information
   - *Systematization,* ordering information according subsystem variables
   - *Transcription,* digitizing
   - *Geo-coding,* locating urban services and potential tourist places.
2. **Analysis:** It uses GIS capacities to calculate relation phenomena between variables.
   - *Concentration,* applying statistical analysis to count amount of units per area.
   - *Proximity,* using buffers to calculate influence areas from punctual, lineal or polygonal attracting sites.
   - *Accessibility,* using network analyst to determinate best routes to interesting sites.
3. **Results:** It assists in the delimitation and qualification of potential sites

   Direct final products
   - Thematic maps by aspect, by screen or ready-to-print
   - Database reports, by search criteria or inventory outputs.

   Indirect sub-products
   - Digital maps, useful as consumables for other applications.
   - Database tables, applicable to interchanges with other external systems.

APPLICATIONS OF IT AND GIS IN TOURISM INDUSTRY

A. Advantage of IT and GIS

Use of GIS technology for tourism planning offers a group of advantages in data documentation and processing (Caldera-Teyes, 1999).

- Thematic maps of variable can be generated in a scales and inexpensive way.

- Information can be updated quickly and easily errors corrected dynamically.
- Results can be shown in cartographic, statistical or table form.
- Information can be integrated easily, through automated relation between different databases that describe the same geographic space according different themes.

These advantages help in the simplification of processing of voluminous geographic-tourist information, such as natural and cultural resources, tourist facilities, accessibility, and transport, land uses among others. Also, they can drive spatial analyses necessary to generate thematic maps or statistical reports, as much for academic, institutional, enterprise interest as well as for tourist requirements. Conformation of collaborative thematic information networks, will allow for handling consistent data closely to link with social realities.

B. Applications of Information Technology

The tourism industry already uses a wide range of IT solutions. Many internal management applications facilitate the management and marketing of tourism organizations. These systems use database for inventory control and administration purposes. The WTO argues, "The key to success lies in the quick identification of consumer needs and in reaching potential clients with comprehensive, personalized and up-to-date information". It also acknowledges that computerized networks, airline Computer Reservation Systems (CRS) have been leading structural changes since early 1970s (WTO, 1994a, 1995d). Tourism operators realized the need for developing interfaces to enable them to communicate effectively with the intermediaries through CRSs. These CRSs were used to manage their inventory and distribute their capacity and to expand their market globally.

The evolution of CRS to Global Distribution Systems (CDS) allowed the managers to expand their geographical and market area by integrating at different levels, that is, with other airline systems and with a range of tourism products. In addition to the GDSs Destination Management Systems (DMSs) emerged to provide integrated solutions of destinations and tourism enterprises and also to amalgamate local systems and providers under one name.

(i) IT for Tourism Demand and Supply Management

The usage of IT in tourism industry is driven by both the development of size and complexity of the market in terms of demand. IT improves the service quality and contributes to higher customer satisfaction.

Consumer benefits of IT will enable greater involvement in planning enabling the niche markets to go. A wide range of opportunities and challenges emerge for even smaller tourism enterprise. At the micro level, tourism planners will maximize their long-term profitability through networking and Internet as these power even the tiny organizations and destinations to be represented in the market place and to network with consumers and partners. Hence, IT changes the best operational practices in the industry rapidly and enables innovative operators to take advantage of the emerging tools.

(ii) Internet

The Internet has also changed the role of tourism intermediaries – travel agencies and tour operators in particular. Travel agencies have been the major brokers of tourism services and the interface of the industry with consumers. IT enables travellers to access reliable and accurate information as well as to undertake reservations at a fraction of time with out much inconvenience through the net from anywhere. Unless intermediaries add value to the tourism product they will gradually lose market share as more consumers begin purchasing tourism products directly from suppliers.

(iii) Information Kiosks

Information kiosks are public access computers, and usually comprise a networked or stand alone computer, and a display screen. Information kiosks are used to deliver information to general public. Information kiosks provide a means of presenting graphical and audible information to individual users. Such kiosks range form the presentation of video presentations through unmodified television monitors, through to fully networked and interactive computer systems.

(iv) Computerized Information Reservation Management Systems and Destination Oriented Systems

One of the latest IT developments in the tourism industry is the proliferation of ***destination-oriented systems.*** Destinations are recognized as the reason for tourism and amalgams of resources, products, facilities and services comprising the 'total tourism product' or the 'travel experience'.

Destination Integrated Computrized Information Reservation Management Systems address the entire range of needs and services required by both tourism enterprises and consumers for specific locations. It becomes evident that these systems would enable destinations and the tourism enterprises to bridge the gap with consumers enhancing inter channel power and the ability to negotiate with intermediaries. They also enable a host population to maximize positive tourism impacts.

(v) Data Access and Interfaces

There is a close relation between the different kind of users preliminary served and access interface to data. There are two ways through which, the system, can be structured to allow data access. The *Public access* is done through systems such Internet/Extranet and the *Local access,* through spatial analysis tools on an Intranet.

Public access implies a restricted use to data, only for its partial visualization according to pre-selected questions. It works as a server-based environment, where clients receive certain outputs through menu selections. Here, users cannot manage directly whole groups of data. This access is possible by Internet or Extranet connection to severs. It uses a generic Internet browser or a map-based browser like ESRI Arc Explorer.

Local access consists to create complex queries, update or generate new data, or create information subsystems about specific data field. It works in a client-server environment, where server supply geo-referenced database to workstations through intranets, and support development of data analysis interface on smart client stations. For this, one may use ESRI Arc View GIS workstations, any of them with Network Analyst extension and other with 3dAnaluyst extension, to create complex thematic maps with concentration or accessibility analysis.

C. GIS Applications

Tourism activity means a highly complex of relations between factors that happens in the host area. As planning and conservation activity, ordering the tourism system implies the management of a lot of information related to the cultural, social and economic context of each reality, and potential relations due to proximity or other location factors. The complexities of developing and implementing tourism plan efficiently and successfully have provided that the current system alone is not sufficient to provide the solution. If tourism resources are to be utilized properly with minimum damage to the environment, decision must be based on more reliable, updated relevant, easily accessible and affordable information.

Geographical Information System for "Development of Tourism Master Plan" has emerged as the technology with considerable potential for achieving dramatic gains in efficiency and productivity due to the distributed nature of tourism related data and need of various types of spatial analysis. Most tourism impacts are spatial—that is the impact of services and facilities are in direct proportion to its geographic proximity to other cultural and natural phenomenon (Pandey, P.K. and Chakraborthy, R).

The following are some of the applications of GIS for an efficient tourism management.

- Generation of a query based system to answer the user's questions.
- Inventorying of existing tourist resources to assess its adequacy.
- To study the terrain characteristics with reference to tourist spots and their interest.
- Studies of Tourism infrastructure.
- Identification of areas of high pressure with a view of finding alternate, less developed areas so as to shift pressure.
- Site Selection, that is, identification of locations for new tourist spots.
- Identification of tourist potentials segments within the tourist spot.
- Type of activities that are taking place in the tourism area.
- Possible impact of tourist spot/activity on socio-economic and environmental aspects at particular location.
- Mapping of the sizes and extent of different habitat types.
- Networking of national parks for efficient management.
- Change Detection Studies.
- Forecast tourism trends, that is, what will be the future state of tourism if the current trend continues over a period of time?
- Assessment of the impact of tourism on environment.

CONCLUSION

This paper clearly demonstrates that tourism marketing is constantly evolving, taking advantage of emerging IT tools to empower tourism marketing and management by providing cost effective tools for organizations and destinations to target appropriate market segments and to develop strategies tools. Therefore a real cost-benefit analysis is required by all tourism industries in order to appreciate their position and best action to take in order to enhance their competitiveness. The development of new IT, particularly the Internet, facilitated the rise of the new tourist who is becoming knowledgeable and is seeking exceptional value for money and time. It is becoming increasingly obvious that to cope up with the increasing tourism demands, the managers have to use the new technologies available. IT provides tools for tourism to face the demands and provide supply to meet the needs and tourism uses IT to increase its needs and to cater to the user requirements through technological developments.

Acknowledgements

We would like to express our gratitude to Dr. N. Sivagnanam Prof. & Head. Dept. of Geography, University of Madras, who has always encouraged and supported us to write research papers. We would also like to thank all the fellow researchers doing extensive work in the fields of GIS and Tourism. Finally we would like to thank the organizers for giving us a chance to voile our thoughts.

REFERENCES

1. Buhalis, D. (2000), "*Trends in Information Technology and Tourism*" in Trends in Outdoor Recreation, Leisure and Tourism (Eds W.C. Gartner and D.W. Lime), CAB International 2000, OUP, UK.
2. *Bukenya. J., (2000), "Application of GIS in Ecotourism Development Decisions: Evidence from the Pearl of Africa*". Research Paper 2012, Natural Resource Economics Programme, West Virginia University Morgantown, WV.
3. Mejia, R., (2000), "*Sigtur-Zulia: An Application of GIS Technologies for Tourism Planning*", ESRI User Conference Proceedings, Accessible at http.//www/.esri.com/library/useconf/proc00/professional/papers/pap709/p709.htm.
4. Pandey, P.K. and Chakraborthy, R. "Geographical Information System (GIS): Information technology for planning management and development of GOA tourism master plan in next millennium", in GIS development. Accessible at http.//www.gisdevelopment.net/Application/Business Geographics.
5. Gupta, A., (2000), "Information Technology and Natural Disaster Management in India", in GIS Development Accessible at http./Avww. gisdevelopment.net/Application/Hazard Mitigation.

SECTION II

APPLICATION OF SPATIAL INFORMATION TECHNOLOGY, GEOMORPHOLOGY, CLIMATOLOGY AND WATER RESOURCES

5

WATER RESOURCES AND HYDROLOGICAL PROBLEMS OF THE LOWER BRAHMAPUTRA BASIN AND ADJACENT AREAS (LBBA), INDIA

S.C. MUKHOPADHYAY

The present paper is concerned with the study on water resources and hydrological problems of the Lower Brahmaputra Basin and its adjacent areas (LBBA) (Fig. 5.1). This author has adopted modern methodology and extensive field work (Mukhopadhyay, 1985, 1986, 2001) in the preparation of this interdisciplinary paper. The given tables (Tables 5.1 to 5.3) present the related aspects of water, its demands and utilization of other resources of the Brahmaputra basin. The annual flow of water recorded near northern Dibrugarh is about 33,000 million cubic metre while it is 455,000 million cu.m. and 510,500 million cu.m. at Guwahati and Dhubri respectively.

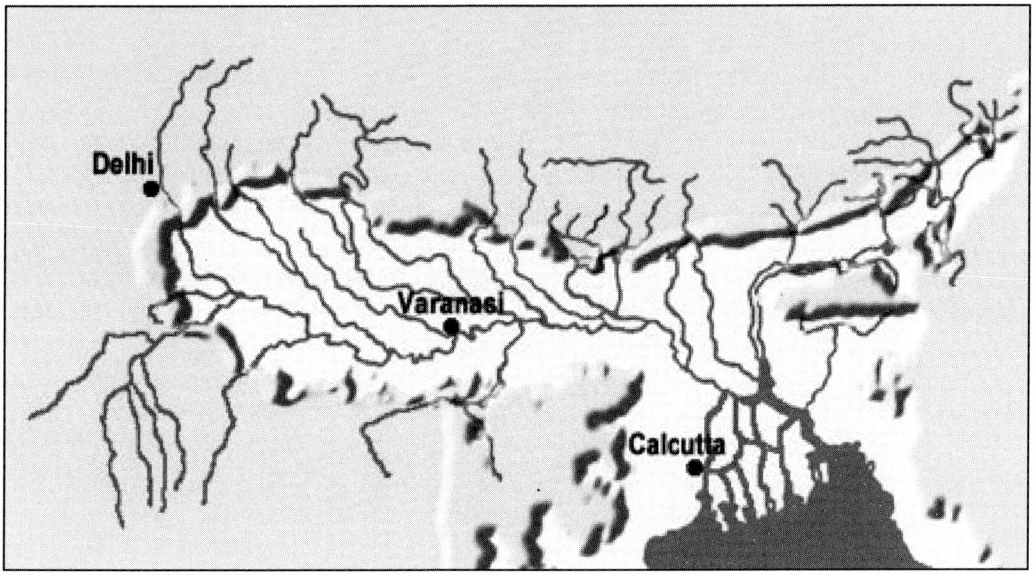

Fig. 5.1: The Location of Lower Brahmaputra and Adjacent Areas
(*Source*: www.google.co.in/images/brahmaputrabasin)

The Brahmaputra rises from the tongue of the Chema Yungdung glacier of the Kailas range (Tibet), China and east of Mansarovar (Konggyo Tsho) Lake at an elevation of more than 5100 m., a.s.l. It is known by the names of Dogzung Zangbo Jiang, Yarloong Zangbo Jiang and Tsangpo throughout its eastward course (following the Indus-Tsang Po Sature F for about 1500 km upto Namcha Barwa (7756 m) where it takes a sharp (boat-hook) bend before entering Arunachal Pradesh (Fig 5.2).

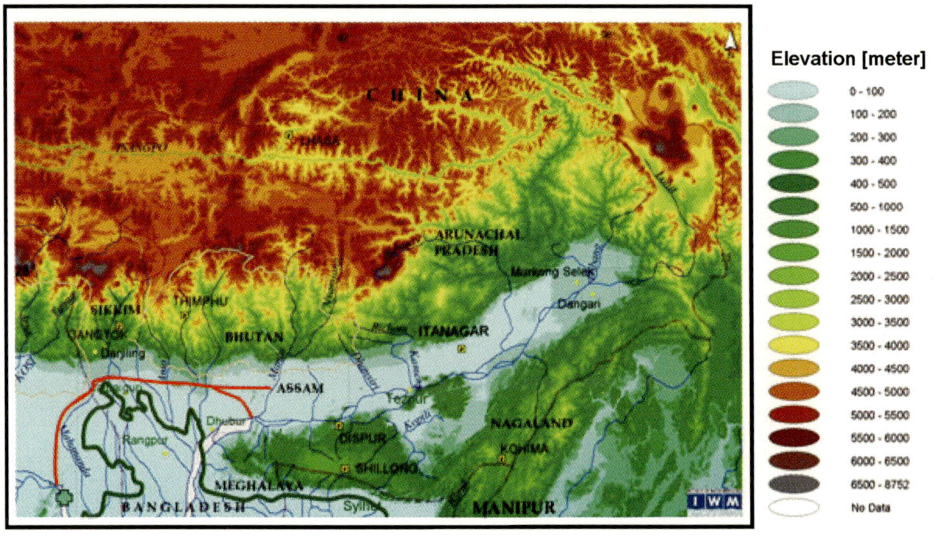

Fig. 5.2: Drainage Map of Brahmaputra Basin
(*Source*: www.google.co.in/images/brahmaputrabasin)

The main flow of the channel is called the Dihang or Siang river in India. Here it is joined by two other tributaries, the Dibang or Sikong and Lohit or Luhit negotiating the Mishmi Upland from the east with the Sadiya Region or Rumkini plain in the eastern extremity of Assam, (27° 44'N to 27° 57'N and 95° 28'E to 95° 54'E; 122.0 m to 165 m. a.s.l.). The combined waters of these three i.e. Dihang, Dibang, and Lohit are known as the Brahmaputra. The total catchments of the river is about 59,000 sq. km. of which 195,000 sq.km. lies in India. From its source of glacier covered Kailash Range (the northern most chain of the Himalaya, it flows eastwards along the depression of the Indus-Tsang Po Sature (ITS) for 1135 km (approximately) extending parallel to the general lie of the Himalayan ranges and then it takes a turn to the northeast near Po (about 3000 m.a.s.l.). Away from the sudden turn at Namcha Barwa (7756 m) further down stream, the river exhibits wide valley carrying V-shaped broad valley (in place of deeper I shaped valley up stream) with formation of river terraces or narrow valley sections where terrace formation is either absent or negligible. The water level in the Brahmaputra is marked by the occurrences of fluctuating flow i.e. variation during the dry and wet season, at Guwahati it is about 9m. Hydrogeomorphological Parameters of the Brahmaputra Basin including population and land use in India are shown in Table 5.1.

Table 5.1: Hydrogeomorphological Parametres of the Brahmaputra Basin, India and Adjacent Countries

1. Total Catchment Area	580,000 km^2
(i) Catchment area within China	293,000 km^2
(ii) Catchment area within India	195,000 km^2
(iii) Catchment area within Bhutan	45,000 km^2
(iv) Catchment area within Bangladesh	47,000 km^2
2. Length from its Source to Confluence with the Ganga	280 km^2
(i) Length within Tibet (China)	1625 km
(ii) Length within India	918 km
(iii) Length within Bangladesh up to confluence with the Ganga	337 km
3. Gradient	
(i) Reach within Tibet	0.00260
(ii) Reach between Indo-China border and Kobo in India	0.00190
(iii) Reach between Kobo and Dhubri	0.0014
(iv) Reach within Bangaldesh	
First 60 kln from India Border	0.00009
Next 106 kln reach	0.00008
Next 92 kln reach	0.00004
Next 79 kln reach	0.00003
4. Discharge Characteristics	
(i) Maximum discharge at Pandu (Assam) on 23.08.62	72794 $m^3 s^{-1}$
(ii) Maximum discharge at Panduon 20.02.68	1757 $m^3 s^{-1}$
(iii) Mean Annual flood discharge at Pandu	51156 $m^3 s^{-1}$
(iv) Mean annual dry season discharge at Pandu	4420 $m^3 s^{-1}$
(v) Mean monsoon flow (June to October)	
Shigatse (Tibet)	507 million m^3
Pasighat (India)	3979 million m^3
(vi) Dischrage per unit area of Basin	
T'sela Ds Zong (China)	0.01 $m^3 s^{-1} km^{-2}$
Pasighat (Arunachal)	0.023 $m^3 s^{-1} km^{-2}$
Pandu (Assam)	0.03 $m^3 s^{-1} km^{-2}$
Bahadurabad (Bangladesh)	0.032 $m^3 s^{-1} km^{-2}$
5. Sediment Load	
(i) Average annual suspended load during flood at Pandu 400 million metric tons	
(ii) Daily mean sediment load during flood at Pandu 2.12 million metric tons	
(iii) Sediment yield	
T'sela D's Zong (China)	100 metric tons km^2
Pasighat (Arunachal)	340 metric tons km^2
Pandu (Assam)	804 metric tons km^2
Bahadurabad (Bangladesh)	1128 metric tons km^2
6. Mean Basin Rainfall (Excluding Bhutan and Tibet)	230 cm
7. Basin Land Use (In India)	114,992.08 km^2
Total forest cover	50,473.84 km^2
Total agricultural land	
8. Basin Population in India	304 million
(Based Goswami 1998)	(143 person/km^2)

WATER DEMAND AND AVAILABILITY IN LBBA (IN INDIAN CONTEXT)

It may be mentioned that further economic development of the study area (LBBA) depends largely on the water resource which fulfils the most basic human need and similar to other areas, is a since qua non for most human activities that fulfil the socio-economic needs of our country. Broadly speaking, the needs have been classified into five sectors like: (a) domestic, (b) irrigation, (c) energy, (d) industrial and (e) others (Table 5.2). Through these multisectoral uses, water ramifies so much into our day-to-day ventures that we have become almost oblivious its great value and have tended take it for granted. Moreover, due to its renewable nature, it has so long been considered as an inexhaustible gift of nature. Present phenomenal rise in population accompanied by the prevailing social and economic need to raise living standards – has necessitated rise in demand of water in every sector of use. Moreover, pollution of various sources of water from several reasons has, on the contrary, brought down the availability of water of adequate quality. In sequel, water has emerged as a scarce commodity in any parts of our country and surroundings, the LBBA is no exception.

Table 5.2: Possible Water Demand in India (2025) and the LBBA (025) with Reference to Previous Years

Purposes	*Annual water demand in the year (M Ha M)*				
	1980	*1990*	*2000*	*2025*	*(LBBA)*
Domestic use	1.6	2.5	3.3	5.2	(0.9)
Irrigation	20.9	46.0	63.0	77.0	(2.0)
Energy	0.9	1.9	2.7	7.1	(0.8)
Industrial use	0.8	1.5	3.0	12.0	(2.3)
Others	2.9	3.3	3.0	3.7	(1.5)
Total	**27.1**	**55.2**	**75.0**	**105.0**	**7.3**

Source: Based on CWC (1995) Aravindan (The Statesman-2002), and computed/surveyed (Mukhopadhyaya 1986).

This note of alarm has necessitated an assessment of the water resources of the country, in order to match the emerging demands of the coming decades with the available supply and to plan future water resources development (WRD) strategies including the concept of Interlinking Draining Basins in North-Eastern India in particular.

Demands

Recent studies reveal that the demand in all sectors of water use was about 55.2 M ha m. in 1990 and 75.0 M ha m in 2000 respectively. Further, water needs in 2025 A.D. are estimated to be 105.0 M ha m. The Table 5.2 may be examined in this respect.

An examination of the table above indicates that irrigation demand is the largest for all the years. In terms of total utilization irrigation demand accounts for 83 per cent in the last decade (1990), 84 per cent (2000) and 73 per cent (2025) respectively. The decrease in percentage in 2025 from that of 2000 is perhaps an indication that irrigation demand has to be restrained from present wasteful use in preference to domestic and industrial demands in future. This fact further accentuates the urgent need for conservation of water resource in the irrigation sector, for an example.

HYDROLOGICAL PROBLEMS

This author has also made an attempt to investigate into the occurrences of major hydro-environmental problems and hazards especially occurrences of 'Floods', and accelerated soil erosion in the Brahmaputra and in its sub-basins like the Kameng (Jia Bharali) in the eastern part. Up streams and Tista along, the western margins (as case study inclusive of design flood both qualitatively and quantitatively (Fig. 5.3).

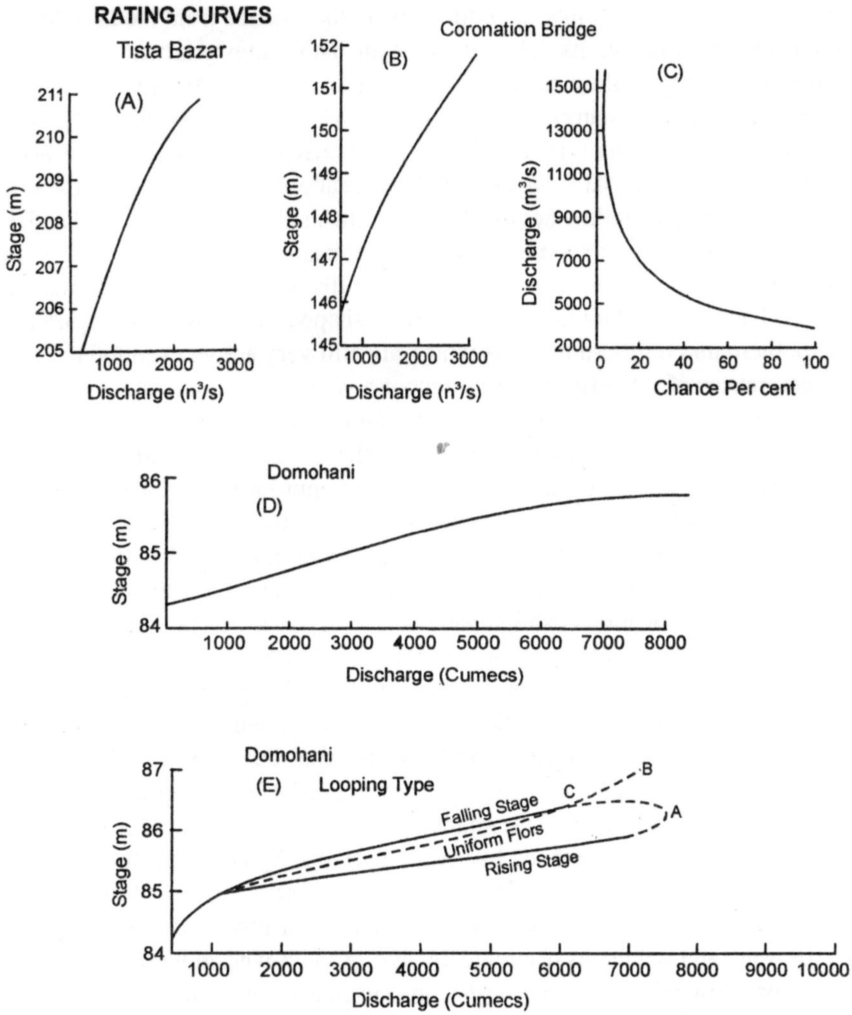

Fig. 5.3: The Steeper Raung or Stage-discharge curves of the Tista at (A): Tista Bazar, (B) Coronation Bridge (1985), and flat rating-curves at Domohani (Jalpaiguri), (D) (1983) including looping type rating curve (E) (1984-85), and Flood-peak-discharge vs. Chance per cent curves (C) (1972-80), ΣQP m³/s = 91.723 by California Method).

IMPORTANT REASONS OF FLOODS IN THE BRAHMAPUTRA BASIN

(1) Very high rainfall (700 mm and more) in a short period i.e. in the South-West Monsoon period (June to September) where monthly rainfall at places cross 1150 mm with records of maximum above 400 m and more of rainfall in 24 hours. (2) Rain storms (cyclones, local severe storms) of one to three or more days duration over the entire BBM basin area e.g. av-storm distribution is 1 in May, 26 in June, 29 in July, 9 in August, 19 in September and 3 in October. (3) The river valleys especially the tributary rivers appear to lack the proper sediment-transfer zone i.e. geomorphologically no intermediate stage. (4) The marked variations in the river valley gradient of the Brahmaputra river, especially for the strength of about 500 km in Dibrugarh to Guwahati is very low in comparison to its distance from the base level of erosion or the sediment sink zone i.e., its delta downstream. This valley gradient of about 16 m/km in the upper-near the gorge-valley section (above Karko 28° 32'N' 95°1'E and Moling at 3075 m.a.s.l.) along the Dihang (Siang). Tsang Po valley (with restricted width and great depth), the boat hook bend near Namcha Barwa, 7755m, a.s.l.) changes abruptly into 0.5 and less m/km beyond the piedmont plain near Pasight, and becomes about 0.1, in average in the lower braided (greater of 4 to 6 km and shallow depth) Part *e.g.* Guwahati downstream. (5) The Brahmaputra (with its braided course) – a tectonically active river basin having very narrow valley section, alluvial gap (flood plain) especially between Garo hills in the South and Bhutan Himalaya in the North. (6) The Brahmaputra flood plain is being shaped by seismic and tectomic movements along its southern and northern margins which are still very active even today.

The occurrences of subsidence of Brahmaputra valley in a comparative sense in respect of the rapid upliftment of its surroundings – *e.g.* (i) the Himalaya in the north (at the rate of 5 to 9 mm in a year), (ii) the Mishmi Massif in the east (at the rate of 9 to 12 mm in a year), and (iii) the Meghalaya Plateau in the south (at the rate of 3 to 5 mm in a year). This author has already presented an interpretative fluvial geomorphological account of the drainage basins of the North-Eastern part of the Indian subcontinent i.e. the Brahmaputra, Ganga, Irradwadedy etc. and their tributary rivers (Sub-basins like the Tista, the Torsa, the Ravi, the Bagmati, the Manipur etc.) Through his earlier publications (Mukhopadhyaya 1982, 1987, 1988, 1989, 1993, 1994, 1995, 1996). In this connection, it seems reasonable to refer to the design flood of one of the important tributaries to the Western Brahmaputra i.e. the Tista river. The given illustrations like Discharge and Stage hydrographs, 1970, 1986 Rating curves at Vhungthang, Coronation bridge, Darjeeling District 1985, and rating curves including Looping type rating curve at Domhani, Jalpaiguri District, 1983, 1985, Discharge *vs* change per cent curve 1972 to 1986 etc. applying California method, Gumble's method (Garg 1978) relating to peak discharge by probability method for near and remote future, and other methods like peak discharge by empirical formulae e.g. Dicken's Formula, Inglis formula, Nawab Jung Bahadur's formula etc. (Varshney 1979), Fuller's formula (Fuller 1957) etc. as presented in the figures. It has been observed that floods or maximum discharge of water and sediments occur mostly during south-west monsoon or rainy season after which the water flow reduces during post-monsoon season i.e. October onwards, and then the minimum is recorded in pre-monsoon season, especially in March or April in a year (followed by pre-monsoonal showers before the burst of monsoon). Mention should be made of some standard formulae (Dicken's Inglis, Fuller's Gumble). For peak discharge by empirical formulae, Varshney 1979. Dicken's formula i.e. $Q = C.A.^{3/4}$ where A is catchment area in km^2, (here are upto

Do-Mohani for the Tista is 9432 km^2) and C is a constant depending upon several factors affecting the peak discharge (11.5, Dey 1990), and the result of required peak discharge of the Tista at Do-Mohani is obtained as 11010 cumecs (m^3 s^{-1}). The peak discharge of the Tista at Do-Mohanis calculated by Inglis' formula (Varshney 1979, Q = 123 A/At 10.4)$^{1/2}$ is 11939 cumecs, by Nawab Jung Bahadur's Formula (Varshney 1979, Q = CA 0.93 – (1/4) log A where C = 48 (Garg 1979) the value is 17751 cumecs, and by Fuller's formula (Fuller 1957), i.e. Qt = CA^{08} (1+0./8 long T) (1+2.66^{-03}), when Qt = magnitude of peak flood having a frequency of t year, C is constant i.e. 2.77, and T is return period in years. Here the magnitude of flood having a frequency of 50 years, Q_{50} = 11572 cumecs, similarly Q_{100} = 12753 cumecs, Q_{200} = 13935 cumecs, Q_{500} = 15496 cumecs etc. The application of California and Gumble's method (Garg 1978) relates to the peak discharge by probability method and at Do-Mohani 1972 to 1987) in near and remote future. The figure of chance per cent curve indicate the magnitude of flood having a frequency of 50 years or a chance per cent of 2 per cent (i.e. 100/50 = 2%) and can be directly obtained from this figure as 13000 cumecs, and similarly the magnitude of floods having a frequency of 100, 200 and 500 years can be obtained as 14000, 15000 and 18000 cumecs (m^3 s^{-1}) respectively. In Gumble's method (Gumbler 1945) $Q_t = Q_p$ + L.s, where Q_p is mean of observed peak discharge i.e. $\Sigma Q_{p/y}$ = 5732.69 cumecs, y is total number of year of floods in record i.e. 16s is standard deviation, i.e. 2191.87 k is frequency factor. Here, for sample site of 16 years (1972 to 1987) the values of k for 50, 100, 200 and 500 years are calculated as 3.293, 2.971, 4.220 and 4.968 respectively. Q_{50} = 5732.96 + 3.293 z 2191.87 = 129509 cumecs, and Q_{100} = 14437 cumecs, Q_{200} = 14983 cumecs, Q_{500} = 16621 cumecs (m^3 s^{-1}) when Q_1 is magnitude of flood having a frequency of t years (Fig. 5.3).

The problem of flood of the LBBA (inclusive the lower parts covering Ganga-Brahmaputra-Methna deltaic plains (Mukhopadhyay 1995) deserves special mention. Because the hydrological, geomorphological, neotectonics and the related geopolitical environments have combined to cause formidable problems in formulating strategies for development especially through water management. In order to find an effective solutions to the flood management problem for the BBM, not only the understanding of the behavioural aspects relating to the knowledge of hydrology of the aforementioned river system is necessary but also the generation and transport of huge sediment (load) is very important. Actually, the amount and nature of sediments affects the entire morphology of the rivers and the water way of the rivers basins available for passing the flood peak get considerably reduced due to sediment deposits, thereby causing flooding of the river banks.

Here, the problems of sedimentation and fluctuating flows are further augmented due to the occurrences of very high degree of Seismo-neotectonics of the undulating terrain of the BBM covering mainly the unstable north-eastern parts of the Indian Sub-continent also under humid environment (Mukhopadhyaya 1996, Nandy 1980 etc.). It may be mentioned that due to recurring earthquakes combined with deforestation and heavy rainfall, numerous hill slides occur which can be observed from satellite imageries and survey maps where several thousands of such scars can be seen. It is necessary to identify all the submarines where the problem of sediment and hill slides are acute. After that necessary rectification measures including afforestation and soil conservation have to be planned. The various methods of flood management have been suggested by National Commission on Floods especially in terms of modifications. The four basic policy issues are: (i) modifying the susceptibility to flood damage, (ii) modifying the flood, (iii) modification of the flood burden, and (iv) bearing the loss.

Table 5.3: Average – Stream Power in Kg/m³/s in monsoon), Allen 1968, seasonal discharge (1975-1991) of water (W in '000 cumecs i.e. ms^{-1}) and sediment (S: in '000 Hectare Metres during Monsoon. Ha M) of the Brahmaputra with Ganga, Barak with Meghna and Manipur with Chindwin etc. basin in the Sink Zones with hints on approx. high range, low range discharge in '000 $m^3 s^{-1}$, etc. (1970-1988).

Rivers	*Stream Power kg/m/s*	*Premon Water Discha.*	*Monsoon (Sediment) m^3/s^{-1} (HaM)*	*Post-Monsoon Oct-Nov*	*High Range (Floods)*	*Low Range (Drought)*	*Runoff ($m^3/s/km^2$) (Monsoon)*	*Soil-Loss (Tonnes/Ha/yr)*
1	*2*	*3*	*4*	*5*	*6*	*7*	*8*	*9*
Dogzung Zangbo	2.68	0.23	2.0 (.95)	1.5	11.5	0.22	3-7	300
Tsang Po	3.99	0.38	3. 6 (1.8)	2.5	19.5	0.35	6-11	500
Brahmaputra	3.44	9.1	38. 3 (3.5)	17.0	28.00	4.50	13-26	750
Ganga	2.97	1.6	30. 0 (1.9)	14.0	24.00	3.15	4-7	600
Meghna	2.73	.56	1.55 (1.8)	.95	6.80	1.15	5-10	400
Subansiri	1.96	.36	3.3 (.56)	.72	8.22	1.35	7-12	600
Kameng	1.87	.35	2.25 (.38)	.55	5.51	1.10	10-14	550
Manas	1.98	.81	5.10 (.74)	1.68	11.00	2.60	6-11	650
Dhansiri	.92	.114	1.12 (.52)	.27	4.10	.65	4-9	350
Barak	1.79	.50	2.00 (.24)	.51	5.00	.95	6-13	400
Manipur	.87	.33	1.36 (.65)	.34	4.00	.70	4-13	350
Raidak-I	.38	.16	2.00 (.17)	.21	.983	.175	7-13	250
Raidak-II	.41	.19	2.50 (.18)	.28	1.802	.35	7-14	200

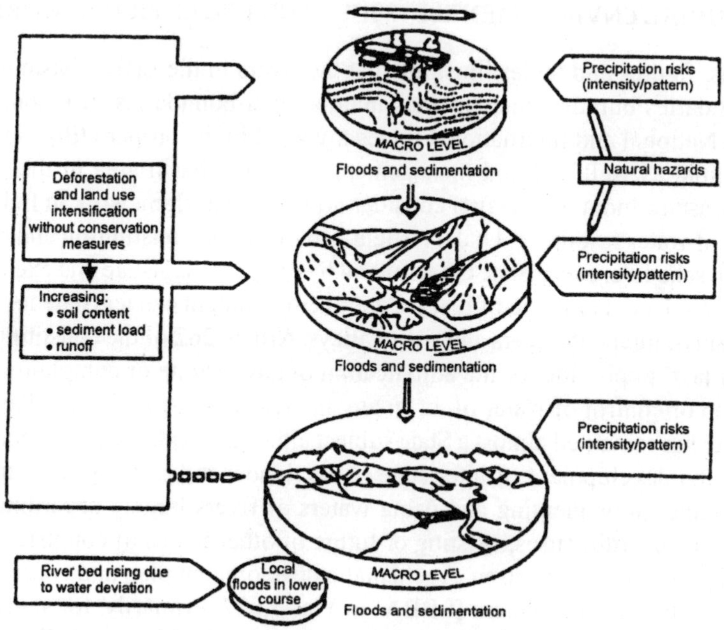

Fig. 5.4: Relationship between Human and Natural Impact Maew-Meso-Micro Levels (Ives *et at.*, Messerli, 1989)

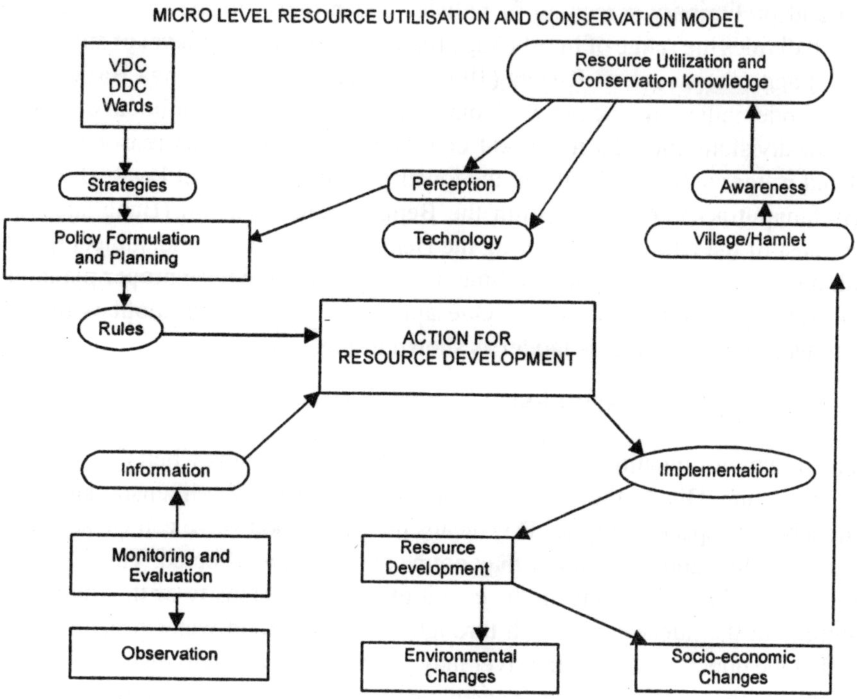

Fig. 5.5: Based on Msserli, 1989 and Klosster, 2002

FLUVIAL ENVIRONMENTAL ISSUES AND LEGAL FRAME WORK

The given maps, figures and tables illustrate that the rivers in the LBBA basin not only cover inter-state boundaries but also encompasses international boundaries. It is thus important to understand the National and International issues involved in the proper utilization of the water resources potential as well as flood management under the basin area in the context of the LBBA, North Eastern India. Let us first consider now the legal framework in India with regard to exploitation of water resources. Under Article 246 (3) of the constitution and entry 17 of list II of the seventh schedule, every State Government has power to legislate and exercises authority in respect of water. Under entry 56 of List L Central Government can legislate for the regulation and development of interstate rivers and river valleys. Article 262 of the constitution authorities Parliament 'by law' to provide for the adjudication of any dispute or complain with respect to use, distribution or control of water of, or in any interestate river or river valley, accordingly. At present water has remained almost a State subject although Central Government can legislate for regulation and development of inter-state rivers. Under the existing legal framework, the States are planning, implementing and using waters of rivers largely according to their own priorities often disregarding uses, existing or future of other States of countries. As a result of number of water disputes have arisen. A schematic development plan and conservation models are shown for the basin management (LBBA) in Yias 5.627. Primarily, for study of such large river basin four hierarchical levels for system studies may be identified. They are: (i) River basin at the National level, (ii) Regional level, (iii) Regional sub-system level, and (iv) Project level. The first level of study seeks to study the next level study by deciding level of regional allocation and preliminary screening of policies and projects in the totality under multiple objectives. It divides the range of interlinkage for analysis in the next level of study. Chaturvedi *et al.* (1975) and others such as Rogers (1995) have analysed the Ganga Basin using a three season L.P. model indicating the net work diagram of the basin with political of states with the geographical divisions into the main and eight sub-basins. It seems reasonable to mention some related water control problems in the lower part of the adjacent lower Brahmaputra i.e. Ganga-Brahmaputra basin areas within the Bengal Deltaic terrain (Bangladesh). Broadly speaking, most of the land areas covering the Bengal Delta is low land subject to tidal action alongwith hurricane surge from time to time. It is important to devise proper planning also for water controls, similar to such areas. One approach could be to adopt with appropriate modifications of the Netherlands model as shown in Fig. 5.6.

RECENT PROBLEMS OF FLOOD AND MANAGEMENT

The basin has a monsoon climate lasting to an average of 4 months when the problem of flood becomes very acute. Due to the very large basin area involved the intensity and frequency of flood varies both in space and time. This means the same flood wave can cause heavy damage in one country while another country may not be significantly affected. This is because of the fact that flood discharge depends on the extent of the area, intensity of the rain storm and the valley storage in the intervening reach through which the flood water propagates. So a very high flood in the upper reaches of the river may get moderate down to insignificant flood in the lower reaches of the worst flood may be generated due to heavy precipitation in lower reaches which the upper reaches may be suffering drought condition.

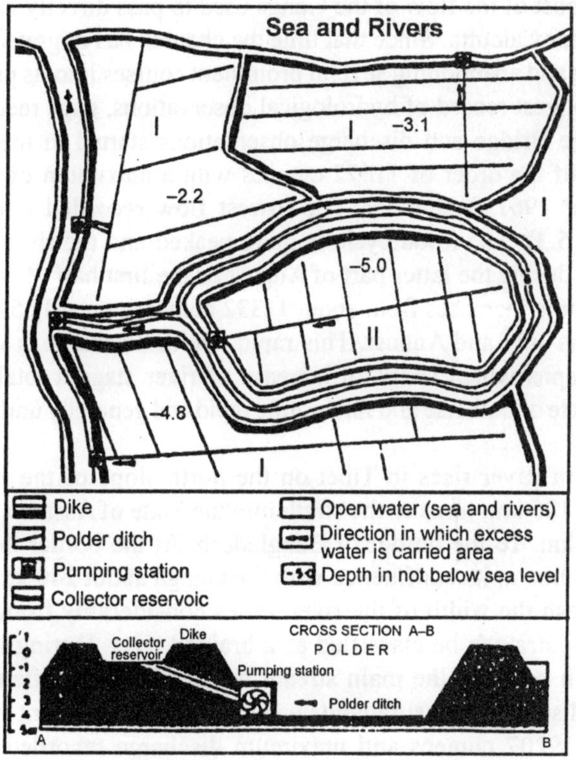

Fig. 5.6: Schematic Diagram Suggesting Water Management: Netherland Model.

Needless to say that the flood causes immense suffering due to loss of domesticated animals, standing agricultural crops, property, communication system, rail, road traffic etc. besides human lives, the flood prone areas have therefore historically suffered severe constraints to their economic development. The region covered by BBM although endowed with plenty of water resources and fertile agricultural lands is economically one of the poorest regions of the World. The region is also amongst the most populous areas of the world and almost ten per cent of total human population occupies about forty million hectares of crop land in the basin. Agriculture has not been developed adequately primarily owing to flooding and drainage congestions. It is also not possible to have industrial development in significant manner because of absence of reliable infrastructural facilities such as roads, communication, electricity, adequate supply of drinking water and water of industrial use and other notable economic activities.

It seems reasonable to discuss some integrated hydrogemorphological aspects on the related environmental hazards also citing examples from the Ganga especially the Lower Ganga river basin in Bangladesh. For the flood management purpose, first aspect required is a study related to hydrology of the river system and the geomorphological changes consequent to very severe floods. The Ganga river rises near Tibet-India border, collects the snowmelt and flows south-west across India, and enters into Bangladesh. It continues south-east across the province until it joins the Brahmaputra river at Aricha, the combined flow being known as Padma river. Prior to the 16th century most of the flow of the Ganga used to pass directly into the Bay of Bengal via. Hooghly river near Calcutta. Since that time the channel has migrated progressively to the north-east occupying and abandoning several prominent courses into its present position. River Ganga has got the longest record of hydrological observations, gage records were initiated in 1910 at the Hardinge Bridge and discharge observations started in the year 1934. Average annual discharge is of the order of 11672 cumecs with a maximum of 80736 cumecs being recorded in the year 1961 September. The lowest flow recorded is a bare 1190 cumecs (Tables 5.1, 5.2 and 5.3). The flood cycle is quite peaked and the river normally reaches its maximum discharge during the latter part of August or the first half of September. During the rising stage of discharge increases from about 11332 cumecs to nearly 56657 cumecs within a period of two months July and August. This rapid rise in discharge is well illustrated in the water level hydrographs which show an increase of river stage a total of 6.5m in the two months period. The rate of both rise and fall is quite rapid and generally uniform (Mukhopadhyay, 1995).

The Brahmaputra River rises in Tibet on the north slope of the Himalayas and flows eastward for about 1125 km. turns to the south into the State of Assam and then turns sharply west for about 600 km. To the border of Bangladesh. At the border the river curves to the south and on this course to its confluence with the Ganga about 240 km North of the Bay of Bengal. In Bangladesh the width of the river varies considerably from about 1 km to more than 13 km Brahmaputra can be classified as a braided river. During low flow the channel shifts back and forth between the main stream banks which are often 6.5 to 13 km apart. Gauge records and discharge measurements were taken up in different years. Annual average discharge is about 19207 cumecs and maximum discharge recorded is 71360 cumecs in 1958 and the lowest flow is about 3286 cumecs in 1960. The discharge hydrograph shows two or three major flood peaks on the Brahmaputra that occur prior to floods in the Ganga. This has got a significant influence on the pattern of river migration below Aricha. Due to flooding of Brahmaputra scourge would occur. However, when the later flood peak of Brahmaputra coincides with the Ganga flood peak the backwater effect caused by the Ganga current probably causes considerably deposition of sediment in the lower Brahmaputra. The Meghna river, for another example, drains the north-east part of Bangladesh and it is predominantly a meandering one, although in several reaches where tributaries join, it shows braided characteristics. The average annual discharge is of the order of 3513 cumecs and flood flow is of the order of 1190 cumecs. It, however, carries heavy sediment load of the order of 20,320 tonnes during a flood. The above three major rivers have dominated the landscape of Bengal Basin (Mukhopadhyay 1995). The sediment brought by them has formed the largest subacerial delta in the world as a larger sub-acucous delta. During flood around 141643 cumecs of water flows in the Bay of Bengal via a single outlet and this causes flooding over a huge tracts of land. It is therefore, evident that Ganga, Brahmaputra and Meghna river-flows principally determine the flooding problem of Bangladesh. The area is subject to moderate floods at every 4 years interval, severe floods at 7 years recurring interval and catastrophic floods at a recurring interval of 50 years.

The Brahmaputra-Barak-Manipur Basin comprises of the largest river basins not in Asia but also in the world which spread over several countries like India, Bangladesh, Tibet (China), Myanmar (Burma) and Bhutan. Hydrogeomorphological and neotectonics considerations are suggestive of the occurrences of considerable instability for the BBM basin area. This elaboration of BBM basin is not only of academic interest may also be utilized for purpose of hydro-planning indicating the latest dynamic characteristics. Any suitable plan for water utilization is not just a function of terrain, slope and precipitation but equal emphasis shall be given to the dynamics of the processes involved in the build-up of the BBM basin itself (Fig. 5.2). From Plate-tectonics point of view the present landscapes including relief and orography with drainage patterns of the BBM has evolved primarily due to active tectonic processes related to northward collision with Tibetan land mass and eastward subduction/collision with Shan-Tenasserim block of the northward moving Indian plate. It seems possible that the tectonic processes are shaping the landscape even today.

CONCLUDING REMARKS

Present author has already discussed the occurrences of water resources and the related hydrological problems of the Lower Brahmaputra Basin areas (Figs. 5.1 to 6, Table 5.1) both in qualitatively and quantitatively providing the available spatial information and data concerned. The LBBA, the Land of Red Rivers and Blue Mountains are endowed with abundant water resources, valuable minerals, vast forest and fertile arable lands supporting both agriculture and sylviculture in addition to the saxicultural activities. Inspite of the presence of these varied and human resources, the area is socio-politico-economically one of the poorest regions of the world. Here, most of the environmental hazards and even the natural disasters (e.g. floods, soil erosion, bank erosion, earthquakes, landslides etc.) found to occur relate primarily to the complex hydro-geomorphological situations of the LBBA in which they take place. These are mainly attributed to the particular impact of water in varied ways thus modifying actively the landscapes especially involving both of physical and cultural landscape patterns as indicated in the text.

REFERENCES

Chatterjee, S.P., (1984), "River and River & Basins", *Ind. Jr. of Landscape System and Eco-Studies*, ILEE, Calcutta, 7: 1-8.

Goswami, D.C., (1985), "Brahmaputra River, Assam, India: Physiography, Basin Denudation and Channel Aggradation", *Water Resource Research*, Am-geophysical Union, USA, 21(7): 959-977.

Guan, Zhihua and Chuan You Chen (1981), "Hydrographic features of the Yarlung Zangbo River", in D.S. Liu (ed.) *Geol. and Eco. Studies of the Qinghai-Xizang Plateau*, Beijing: Sc. Press, 2: 1693-1704.

Mukhopadhyay, S.C., (1993), "Fluvial Environment of Meechi-Sankosh Interfluves", *Ind. Jr. Landscapes Sys. And Ecostudies*, Calcutta., 16 (2).

Mukhopadhyay, S.C., (1995), "Fluvial Systems of the North and Eastern Indian Subcontinent with special reference to water use in the sediment sink zones", *Ind. Jr. of Landscape Systems and Eco-studies*, ILEE, Calcutta, 18(2): 170-206.

Nandy, D.R., (1980), "Tectonic Patterns in North Eastern Indian", *Indian Jr. of Earth Sciences*, Calcutta, 7(1): 103-107; and 1981: "A Discussion" *Indian Jr. of Earth Sciences,* Calcutta, 8 (1): 82-86.

6

RECENT CHANNEL DYNAMISM: A COMPARATIVE STUDY OF GANGA AND BRAHMAPUTRA RIVER SYSTEMS USING REMOTE SENSING GIS INTEGRATION

V.A.V. RAMAN AND S.K. PAL

INTRODUCTION

The *Geographical Information System*, **(GIS)** and the *Digital Image Processing,* **(DIP)** are techniques applied through a computer system, the main components of which are *hardware* (the physical parts of computer along with its associated peripherals, like scanner, plotter, digitizer and printer), *software* (the computer programmes. Packages that run on the computer) and the appropriate *procedures* (techniques for the project implementation). They are designed to handle spatial data in vector and raster data formats, the latter particularly applicable where remotely sensed images are used and DIP is associated with it.

In this study GIS is applied to integrate data layers i.e. the river channel configuration in four different years, to study the temporal change in channel planform of Ganga river between Patna to Mokama in Bihar, India.

On the other hand, the *function of DIP* is to perform appropriate methods of processing to enhance the multi-spectal digital data as per the user requirement that improves the visual interpretation to extract specific information. In the process of registering the satellite image of a part of river Brahmaputra around Kaziranga with the base map (S.O.I. topographical sheet), two-time period database could be generated at the same scale that helped in their digital overlay. Based on this the river channel configuration and the planform change between the two database is studied.

Integrating data through GIS and Image Processing for change detection

The ability to integrate data from more than one source using map overlay technique is perhaps one of the key *GIS analysis* function (*Heywood, et. al.*, 2000).[1] Using GIS it is possible to take different thematic map layers of the same area and overlay them, one with the other to form a new layer. This technique is applied to integrate the maps of channel planform of Ganga river.

By overlaying these theme maps, it has been possible to observe the variation in the river's bankline and the channel pattern and decipher the temporal changes. *Map overlay* has many applications. On one side, it can be used for the visual comparison of the data layers or the layers can be merged with either of them to provide new spatial data set as an output layer and be analysed for change.

Image processing methods are grouped into three functional categories for all practical purposes namely *image restoration, image enhancement* and *information extraction*. All multi-spectral digital data from one single imaging device have the same format. The *Digital Number* (DN) from one band for a particular *pixel* along a particular line has the same spatial coordinate for DN in all other bands relating to the same sensor/surface. The various combinations of the bands are displayed through a monitor or digitally written/plotted to produce a high-quality colour picture, most commonly displayed as a False Colour Composite, FCC image. The channel bankline configuration from the image and the corresponding Survey of India, S.O.I. topographical sheet are created as *vector layers* and are integrated into overlays for analysing the banking shifts.

The GIS applied through *AutoCAD Map 2000, ARC/Info workstation 7.2.1 on window NT 4.0, Mapinfo 5.0* and Image Processing with *PCI's GCP Works* and *Image Works version 7.0.0,* part of the integrated set of programmes in the PCI EASI/PACE software system.

DATABASE AND METHODS

In this analysis, the work required preparation of maps showing the river channel configuration in different years. It is possible to generate these from the following topographical sheets and satellite imageries (Table 6.1).

Table 6.1

Year	*Source Details*	*Reference No.*	*Specific Date*	*Scale*
Ganga River Channel				
1950	Survey of India topo, sheet	72 C	–	1:254,440
1974*	Survey of India topo sheet	72 C	–	1:250,000
1992	Satellite Imagery-FCC	IRS-1B 21-50 (LISS-I)	7th November.	1:250,000 approx.
2001	Satellite Imagery-FCC	IRS-1b 21-50 L2 A1 (LISS-II)	16th January	1:125,000 approx.
Brahmaputra River Channel				
1977	Survey of India topo sheet	83 F	–	1: 125,000
2001	Satellite Image (digital) IRS data NRSA	IRS-1B 15-49 L2 B1 (LISS-II) Band, 1,2,3,4 Standard product	10th January	–

* The database was improved with information from Aerial-Photo Mosaic and its visual analysis.

In this study GIS as a tool is applied for change detection and bankline migration of Ganga river channel between Patna to Mokama in Bihar in past 50 years from 1950 to 2001. All details regarding river channel planform, sandbars and alluvial islands are traced into thematic map from each of the database. The images were visually analysed and the required details were traced out. The railway line and the station network, along with the two important bridges are also marked in one of the layer taken from SOI topographical sheet. Four GCP's are selected and transferred in all maps for registration.

The application of GIS in the study is categorized under the following sequential steps of progression: *Database design, Database automation, Analysis* – through data integration (overlay and query), and final *Presentation* in thematic maps with the analytical results. The process involved converting the map layers along with its features into a digital vector format on the computer system. All map layers are scanned for conversion into raster format and then digitized, edited and rebuilt. To impose a scale factor, it is necessary to convert them to the real-world geo-referenced system, a process known as 'transformation'. All the map layers/ coverage are then rebuild to their feature topology and the attribute data joined with the database. Once brought into the computer framework, the data layers are integrated for analysis. A GIS enhances this process by providing tools, which can combine the data layers in meaningful sequences of relationships within and the between the data sets, thus increasing understanding of the real world. Once the digital database is ready, polygon overlays are applied and queried with different modes of command expression depending on the requirements of the study.

The channel planform change in the Brahmaputra river along the Kaziranga National Park between 1977 and 2001 is attempted through the image processing technique. The IRS LISS 2 digital data supplied in a 7 bit data type requires the file format to be changed as a pre-processing and is applied, so that the processing of the database could be undertaken. The *image processing* work being different from the GIS approach said earlier is based on the following steps of progression: *Configuring the image, Image registration, Creation of vectors* i.e., maps of channel planform, *Map overlays* for change detection.

The whole process is applied through the above mentioned PCI's EASI/PACE Image Works and the image is first enhanced for better visual interpretability. Further, both the image and map are registered and geo-referenced to bring both to the same scale and planimetric orientation, within the DIP mode thorough the GCP Works software. Once the map and image are geo-referenced, a subset is created for the purpose of analysis. Separate vector layers are created from both the image and map, in which the respective river channel configuration are digitized, and then overlay analysis is performed.

PATTERNS OF CHANGE

Channel shifting or channel planform change is one of the most conspicuous forms of the geomorphological changes in a fluvial landscape. Changes in the form often appear to be observable and measurements of channel migrations, which have been attempted by many like *Goswami* (1992)[2] *Goswami, et. al.* (1993)[3] and *Leys* and *Werritty* (1994)[4] and others in different ways.

The morphology and the behaviour of the river channel make changes in response to the variations in the flow regime and pattern of sedimentation. In alluvial plains, two types of movements in channel changes can be observed based on changes in the channel width and/or total lateral shift of the channel. Generally, in a meandering channel, the change is associated to the meander loop expansion and cutoff by migration and avulsion, whereas in a braided river the total lateral shift of the banklines are observed.

CHANNEL PLANFORM CHANGE: THE CASE FOR GANGA RIVER

The present study is attempted in an aerial length of 108 km along the Ganga river between Danapur (west of Patna) to Mokama, as per the database of two Survey of India maps in its edition of 1950 and on 1974. For the two satellite images, their 'November 1992' and 'January 2001' versions are compared (Table 6.1). It is kept in mind that the images are of the normal winter level of the water flow so that the channels can be compared to its counter parts from the topographical sheets. The satellite reference of the 1992 channel is exactly from the Patna City to Mokama covering a distance of 95 km and for the 2001 it is again from Danapur, but only until Barh covering an aerial length of 78 km.

The comparative analysis of the planform changes between the four years database is shown in figures 6.1 to 6.4. The changes are observed with reference to the area on to which the river channel has 'transgressed' (i.e., the channel shifted on to the non-channel area beyond the bankline). In contrary the areas from which the channel has withdrawn towards itself is defined as 'regressed'. The areas common to the channels are designated as zones of 'no change'.

Visually the graphic output does show the changes occurred within the planform, the Table 6.2 provides the area wise data of the changes in the channel planform. It is observed that only between 1974 and 1992 the channel has regressed more than the amount of area it transgressed.

Table 6.2: Area-Wise Change in Channel Planform of Ganga River Channel Between Patna to Mokama 1950 to 2001

Channel Change	*1950-1974*	*Percentage*	*1974-1992*	*Percentage*	*1992-2001*	*Percentage*	*1950-2001*	*Percentage*
1. No Change	711.22	73.49	677.95	79.90	557.37	89.02	605.46	76.41
2. Transgressed	194.01	20.05	79.70	9.39	49.47	7.90	145.19	18.32
3. Regressed	62.55	6.46	90.81	10.70	19.25	3.08	41.77	5.27
Total Area between the banklines	967.78	100.00	848.46	100.00	626.00	100.00	792.42	100.00
Total Change (2+3)	256.56	26.51	170.51	20.09	68.72	10.97	186.96	23.59
Difference (2-3)	131.46	13.58	-11.11	1.31	30.22	4.83	103.42	13.05

Values in Sq. Km

Planform change in 1950-1974: Between 1950 and 1974 the Ganga river channel shifted significantly northward and in an area northwest of Patna, it captured a major part of a tributary of Gandak, Mahi Nadi on its northern bank. Near Hajipur, along the alignment of Mahatma Gandhi Setu (Patna Hajipur road bridge), the Gandak river straightened its course and major part of its discharge was transferred into this channel. The Ganga channel transgressed and regressed along both of its banks around Raghopur Diara to Barh. The Ganges created a major meandering bend near Bachwara and transgressed on to the non-channel areas from Bachwara to Mokama along the northern bank. The total changes in 1950-74 by the processes of transgression (+2) and regression (–) was measured as 26.51 per cent of the total channel area, of which the transgressed area was 20.05 per cent while the channel withdrawal was by remaining 6.46 per cent (Figure 6.1).

Planform change in 1974-1992: In this period out of the total change of around 20 per cent, transgression occurred in 9.39 per cent of the total area, whereas regression exceeded to 10.70 per cent. This is significantly attributed to the channel straightening its course from its last deflection near Bachwara (Figure 6.2). A total area of 1.31 per cent from the channel area got added to the non-channel area.

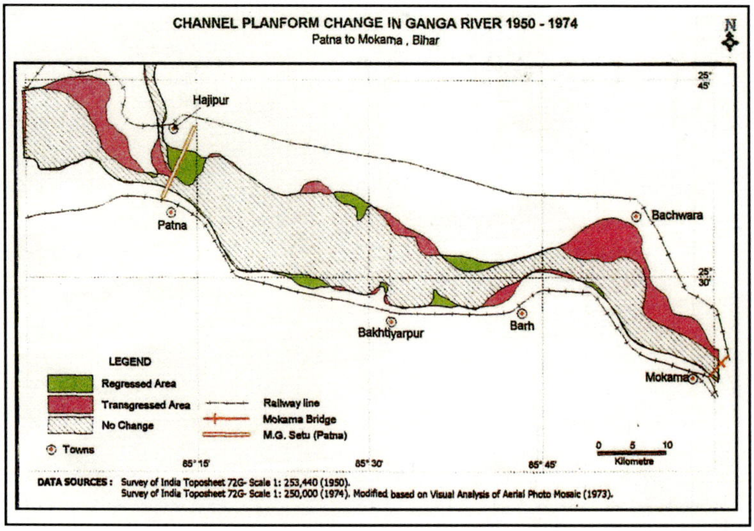

Fig. 6.1

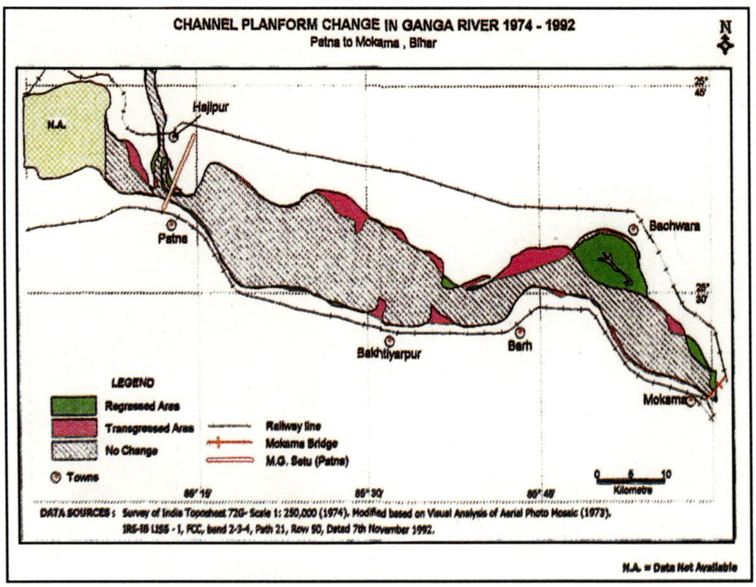

Fig. 6.2

Planform change in 1991-2001: During this period only about one-tenth (10.97%) of the area underwent changes, within which almost eight per cent was transgressed from the non-channel areas within the embankment. The area of 4.83 per cent was added into the channel from the non-channel areas. The coverage area in this time period 1992-2001, was found to be (626.09 sq. km.) least compared to the other periods, because data pertaining to only 65 km of aerial length was available, as the 2001 data is from a hard print satellite imagery at the scale of 1:125,000 (Figure 6.3).

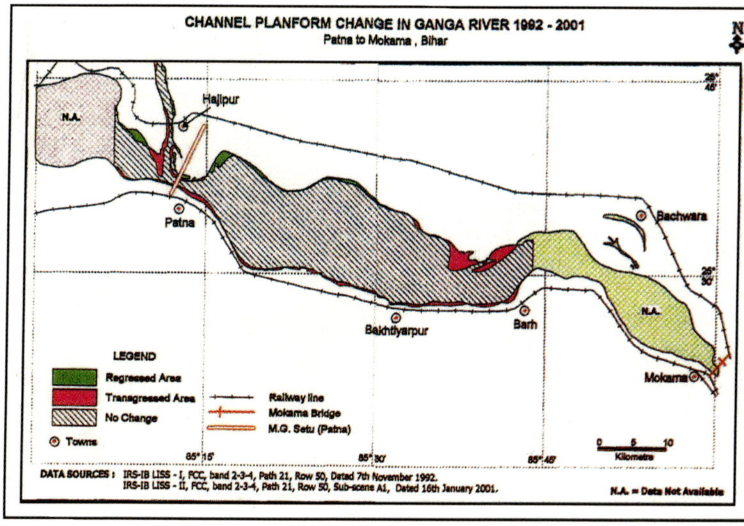

Fig. 6.3

Planform change in 1950-2001: In this the channel change in a span of 50 years is taken up by comparing the planform of 1950 and 2001. It is observed that it is not much different from that of 1950-74 concluding to the fact that most of the changes occurred in the first half of the time scale taken into study. The channel has transgressed on both sides of the bank around Barh as it has widened the channel width. It is observed that in the total covered area of 792.42 sq. km, the channel area has increased by 13.05 per cent (103.42 sq. km.) over the five decades between Danapur in the west of Patna to Barh in its east. In other words, the total amount of area eroded and taken over by the channel during the period was of 103 sq. km. (Figure 6.4).

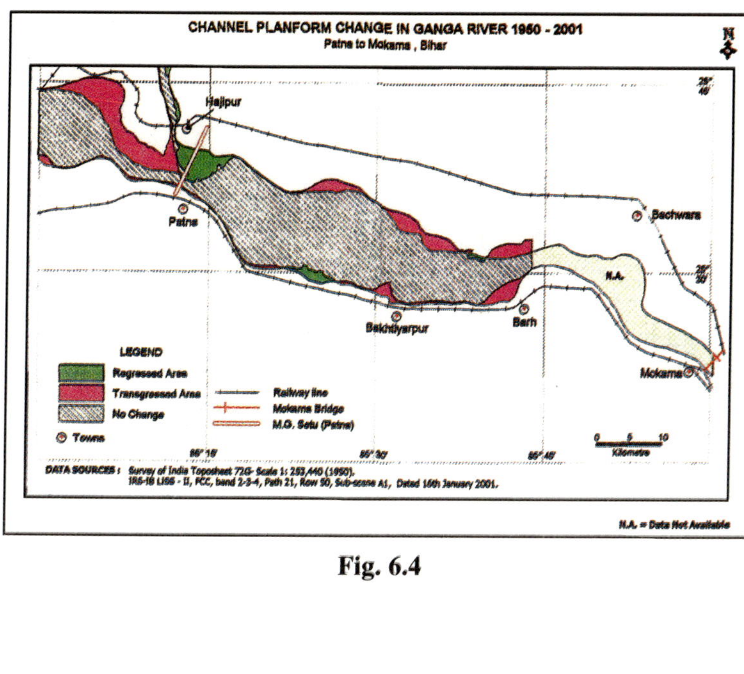

Fig. 6.4

During the entire period the channel planform has changed due to fluctuation of channel form and in those areas where the bankline on either side has not changed much could be considered stable. Maximum shift between the banklines in transgression is found to be 5.7 to 6.4 km and in regression it is about 4.8 km. The river depicts an irregular meandering pattern in this stretch as observed from the multi-channel form taken by the active channels within the channel belt during lean period, the same during wet season in a bankfull stage may depict a large meandering channel. The active channels within the banklines reveals that they continuously get displaced and occupy different positions at different times and may change their courses within the bank limits when the water level recedes after every bankful discharge when scouring will be high. The alterations on the sand bars and alluvial islands also may be due to similar reasons. It is significant to note that the northern bank is unstable comparatively to the southern as the transgression is observed mainly in two places, where the river has created a major meander bend and has modified the same in successive periods, showing its dynamic character.

CHANNEL PLANFORM CHANGE: THE CASE FOR BRAHMPAUTRA RIVER

The Brahmaputra river with multiple channels moving around mid-channel braid bars called *chars* is primarily a braided river, the bankline of which is unstable. The study of the two time database of 1977 and 2001 has provided the basis for understanding the instability of the planform through bankline migration of the river and the subsequent variations in its channel width.

In order to find out the nature and amount of bankline shift and to assess the channel planform change the topographical map with respect to 1977 edition and 2001 satellite imagery: they are co-registered and brought to the same scale by the principle stated earlier. The study pertains to an aerial distance of 75 km downstream from the point where the river Subansiri has its confluence with the Brahmaputra river near Pichalamiri in its north bank. The Brahmaputra is also joined in its southern bank by Dhansiri river next to the eastern end of the Kaziranga Park. Downstream the river course extends until Vishwanath, where a spill channel of the river Garabagi Suti takes off from the parent river.

Lateral shift of the channel is a common index for studying the channel planform change. Channel movement is essentially a vector aspect, with channel moving both laterally and downstream through processes like bend migration, bank erosion and avulsion. Identifying few transects, which are representative of lateral shift, lines are drawn perpendicular to the flow across the river channel (Figure 6.5). These transect positions are used for the analysis of lateral shift in the river planform.

The bankline migration studied is based on five transects (or segment 1 to 5) along which the shift is measured. Two sets of information are collected, (one) the bankline shift along the transect, and (two) to study the changes in channel width over the period of the study.

Tables 6.3 and 6.4 shows the details of the measurements taken directly in an automated form through the software analysis. Along the *1st transect*, the northern bank with respect to 1977 has shifted 365 metres southward by 2001. Similarly the south bank also has shifted towards south by 1.73 km. As there is a shift of both the banks towards south the average difference between the channel width is of 1.36 km of increase.

Table 6.3: Bankline Migration of Brahmaputra River at Selected Cross-Sections Around Kaziranga National Park (1977-2001)

Transect	*A-A'*	*B-B'*	*Change of Bankline*		*Shift in Bankline*
	North Bank	*South Bank*	*North Bank*	*South Bank*	
1.	365 m (0.36 km)	1733 m (1.73 km)	Regression	Transgression	Southward shift of both bankline
2.	No change	3617 m (3.61 km)	No change	Transgression	Southward shift of southern bankline
3.	992 m (0.99 km)	2003 m (2.00 km)	Transgression	Regression	Northward shift of both bankline
4.	1470 m (1.47 km)	2842 m (2.84 km)	Transgression	Transgression	Channel widening on both sides
5.	3071 m (3.07 km)	1993 m (1.99 km)	Transgression	Regression	Northward shift of both bankline

Table 6.4: Change in Channel Width of Brahmaputra River at Selected Cross-Sections Around Kaziranga National Park (1977-2001)

Transect	*Channel Width as on S.O.I. Toposheet (1977)*	*Channel width as on Satellite Imagery (2001)*	*Difference*	*Change in Channel Width*
1.	7356 m (7.35 km)	8724 m (8.72 km)	+ 1.368 m (1.37 km)	Increase (-do-)
2.	6348 m (6.34 km)	9965 m (9.96 km)	+ 3617 m (3.62 km)	Increase (widen)
3.	5563 m (5.56 km)	4552 m (4.55 km)	-1011 m (1.01 km)	Decrease (narrowed)
4.	5320 m (5.32 km)	9632 m (9.63 km)	+ 4312 m (4.31 km)	Increase (-do-)
5.	4985 m (4.98 km)	6063 m (6.606 km)	+ 1078 m (1.08 km)	Increase (widen)

Values in Sq. Km.

The *2nd transect* shows no change in the north bank, but substantial shift towards south of 3.6 km by the south bank. The channel has widened by the same amount and is now almost 10 km (9965 m). Along the *3rd transect*, the channel has transgressed to the north by 992 metres and the southern bank has shifted northward by 2 km. This shows that the channel itself has made a northward lateral shift but overall the channel width has narrowed by over 1 km.

The significant change is of the channel widening of almost 10 km with an increase of 4.3 km along the *4th transect*. The *5th transect* also show an increase in the channel width to about 1 km (1078 m) and the river migrates northward by 3.71 km. In comparison to the south bank shift of 2 km. (1993 m) northward.

Thus there has been a substantial increase in the channel width as the river shifts laterally within the study area. The tansects 2, 3 and 4 show the changes occurring along the Kaziranga National Park, wherein from both the sides the river has cut into the area, reducing its stretch, whereas the river has aggraded the park in its central area towards the river. The average rate of

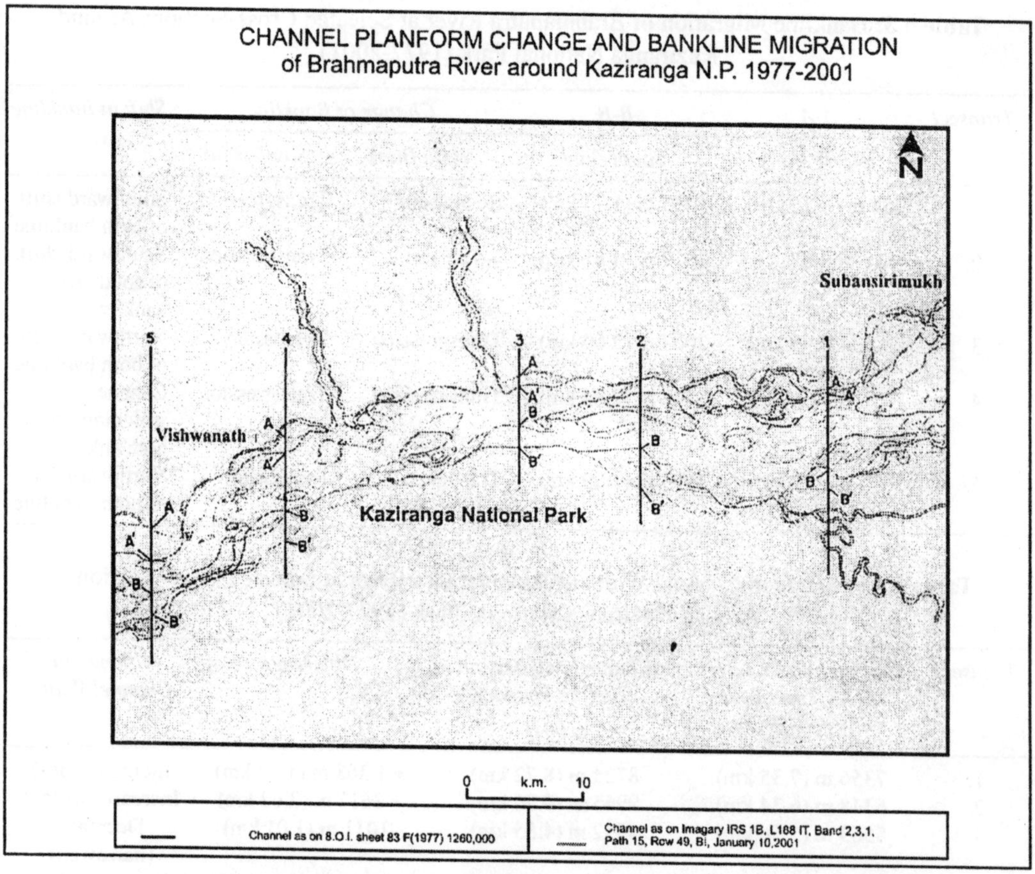

Fig. 6.5

lateral shift in the bankline of the north bank is 1.18 km compared to that of south bank of 2.44 km along the reach. The Kaziranga area is a part of the active flood plain of the river as numerous spill channels are present to carry water during every bankfull discharge. The confluence of Subansiri river with Brahmaputra exerts an influence in making the river migrate laterally southward. In the southern bank along transects, 1 and 5 embankment protection measures are present, which makes the river erode the banks in areas devoid of such controls, and Kaziranga area (whose position can be seen along the tansects 2 to 4 on Figure 6.5) is the one that is not protected from the fury of the river. Observing the overlay also shows, the multitude of sandbars in the channel and their shift by location.

To sum it up, the outward (transgression) migration of both the banklines as in transect 4 causes a marked increase in channel width. Greater outward and lesser inward (regression) migration of the banklines as in transect – 1 and 5 also causes channel widening. On the other hand greater inward and lesser outward migration of both bankline reduces the channel width as in transect 3. Lastly, only transgression and no inward migration as of transect 2, will only lead to widening of the channel. The overall channel planform of the Brahmaputra river in this stretch shows more area being transgressed and the river ends up getting widened. The possible reasons for the widening of the channel are from its response to bar development and development of 'anabranches' along the river channel.

CONCLUSION

This work was an attempt to integrate the two techniques: Remote Sensing and GIS to study the recent river channel dynamics and the output has provided a set of graphic overlays and reliable sets of data showing the channel pattern change and bankline migration. The study shows that the two river channels, Ganga and Brahmaputra undoubtedly have made significant shifts causing change in their channel configurations. In comparison, the Brahmaputra is extremely unstable for most part of its channel banks by its total channel shifts whereas the same is not true for Ganga, which shows to an extent, a degree of stability in its southern bank.

REFERENCES

Heywood, I., Sarah, C. and Carver, S., (1998), *An Introduction to Geographical Information Systems*, Addison Wesley Longman Limited, England.

Goswami, U., Sarma, J.N. and Patgiri, A.N., (1999), *River Channel Changes of the Subansiri in Assam, India, Geomorphology*, 30: 227-44.

Leys, K.F. and Werritty, A., (1999), *River Channel Planform Change: Software for Historical Analysis. Geomorphology*, 29: 107-20.

7

STUDY OF CHANGES IN PRE-MONSOON CONVECTIVE DEVELOPMENTS OVER GANGETIC WEST BENGAL

U.K. De and S.C. Mukhopadhyay

INTRODUCTION

In India convective developments over Gangetic Plains, is an important weather phenomenon during hot pre-monsoon period (March—May). Incidentally pre-monsoon thunderstorms are the only major source of rainfall during March to May in this region. So the study of behavioural pattern of convective developments over Gangetic Plains has great relevance for hydrological management in the pre-monsoon months. Since the meteorological observatories are not dense here, the study through weather radar is to identify the convective prone zones around DumDum (22.38° N', 88.28° E) (Fig. 7.1), which might be useful for mesoscale study. A comparison is made whether convective developments as well as the depth of the thundercloud are on the decline in recent years.

DATA AND METHODOLOGY

The study is based on weather radar observations for a period of 10 years during 1980-89. The 10 year period is split up into two 5-years periods of 1980-84 and 1985-89 to see whether any change is occurring in convective activities in recent times. An area coverage of 300 km by radar around DumDum is divided into 60 10n8S, each having radial width of 60 km and substaining 30° angle at DumDum. The number of developments for the two periods 1980-84 and 1985-89 for the entire pre-monsoon period is presented in Fig. 7.2. The maximum and minimum convective zones for all developments around DumDum are shown in Fig. 7.3. Next, the matured developments (when the cells are no longer growing) or the two periods are presented in Fig. 7.4. The zones having most and least matured developments around DumDum are shown in Fig. 7.5.

Fig. 7.1: Location of the Study Area

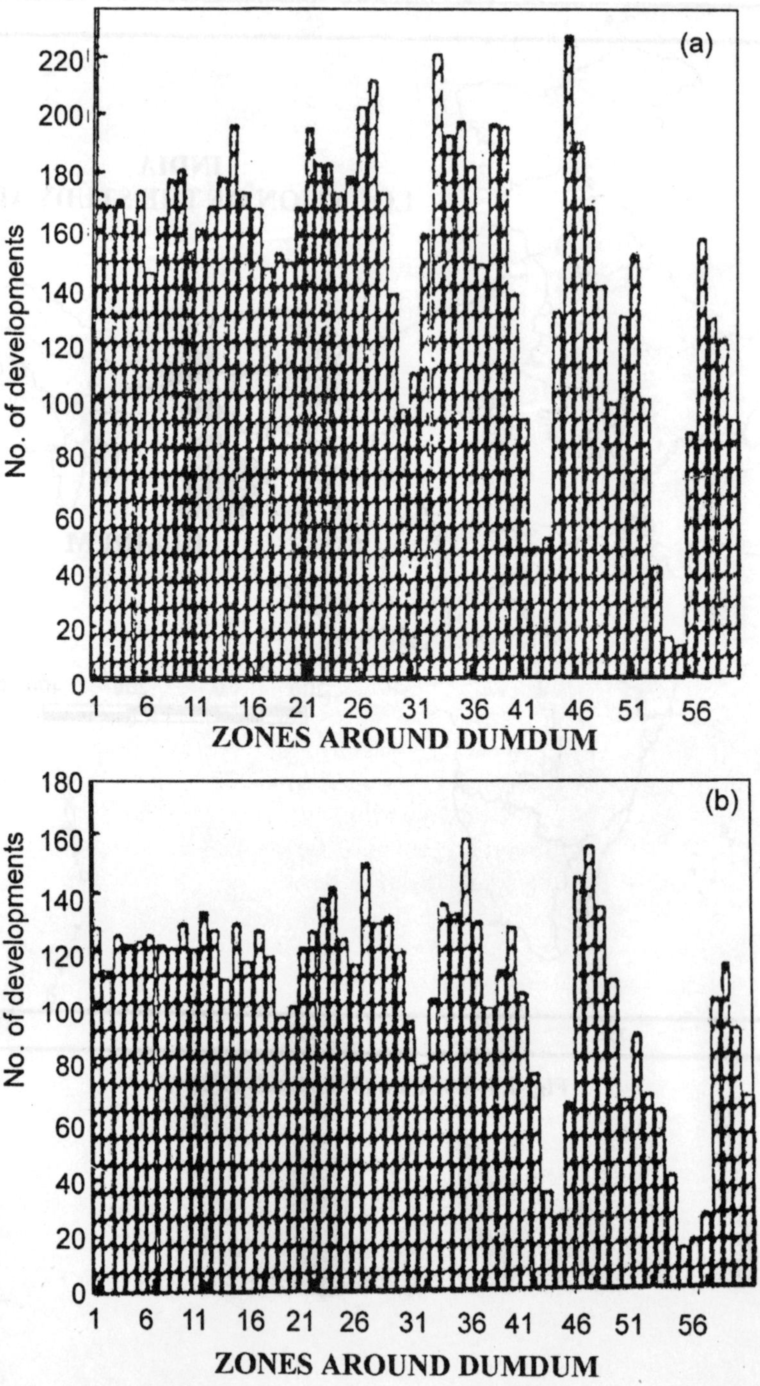

Fig. 7.2: No. of Developments of Convective Developments for (a) Pre-monsoon Period During 1980-84 and (b) Pre-monsoon Period During 1985-89.

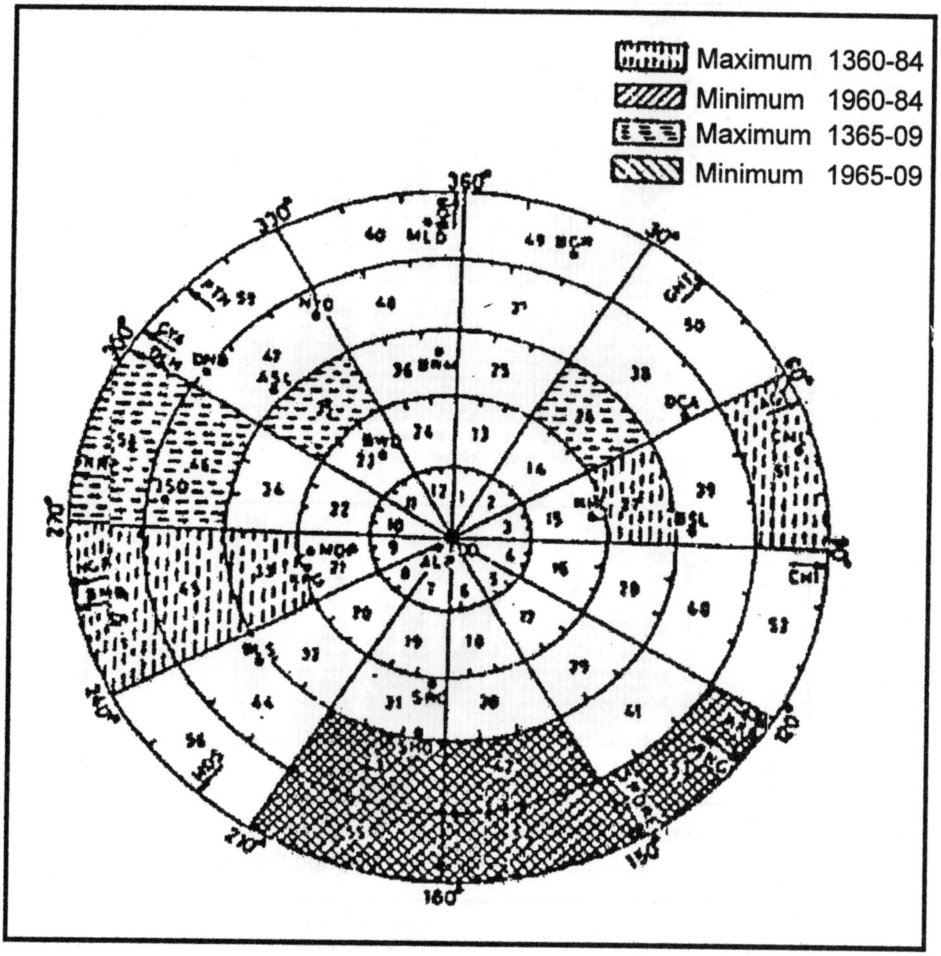

Fig. 7.3: Most and Least Convective Zones around DumDum

RESULTS AND DISCUSSIONS

The number of convective developments increases significantly from March to May for both the periods. However, in the second period (1985-89) developments decrease compared to the first one (1980-84) in each of the three months (Fig. 7.2). In the first period the maximum developed zones are towards south of west and north of east around DumDum, whereas in the second period the corresponding zones have shifted slightly northward (Fig. 7.3). The number of matured developments increases from March to May for both periods. Again those have declined in the period 1985-89 (Fig. 7.4). In comparison to the period 1980-84 the maximum development zones have shifted slightly northward during 1986-89 (Fig. 7.5). The frequent cloud height was observed to be 8 km during the first five-year period (1980-84) and this has declined to 6 km in the second five-year period (1984-89) (Fig. 7.6 and Fig. 7.7).

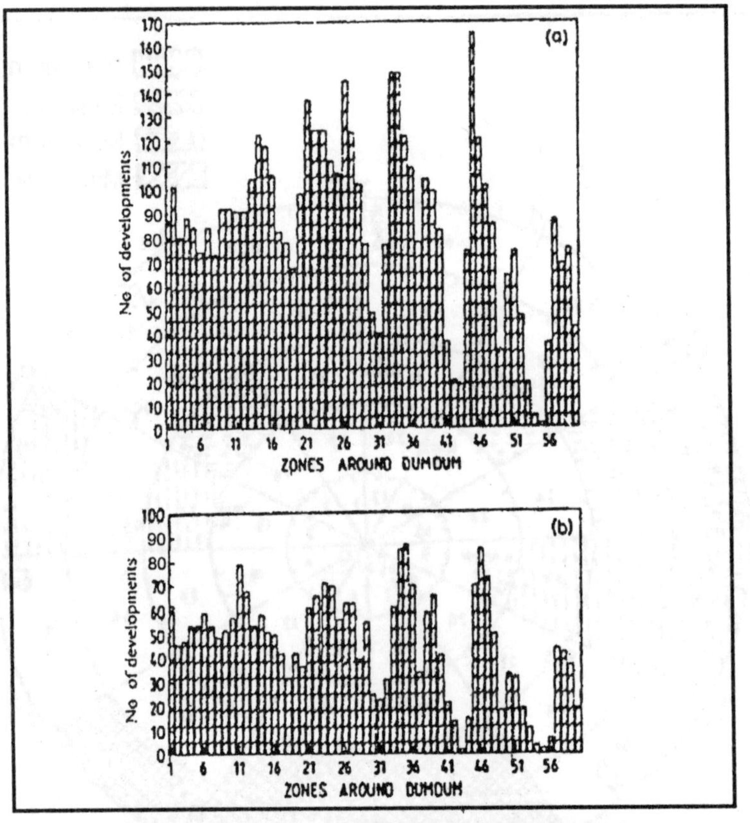

Fig. 7.4: No. of Developments of Matured Developments for (a) Entire Pre-monsoon Period 1980-84 and (b) Entire Pre-monsoon Period During 1985-89.

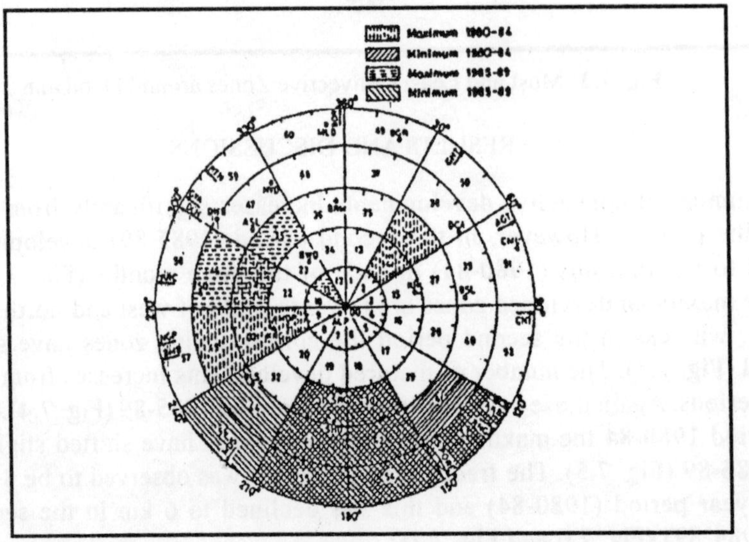

Fig. 7.5: Zone Having Most and Least Matured Developments.

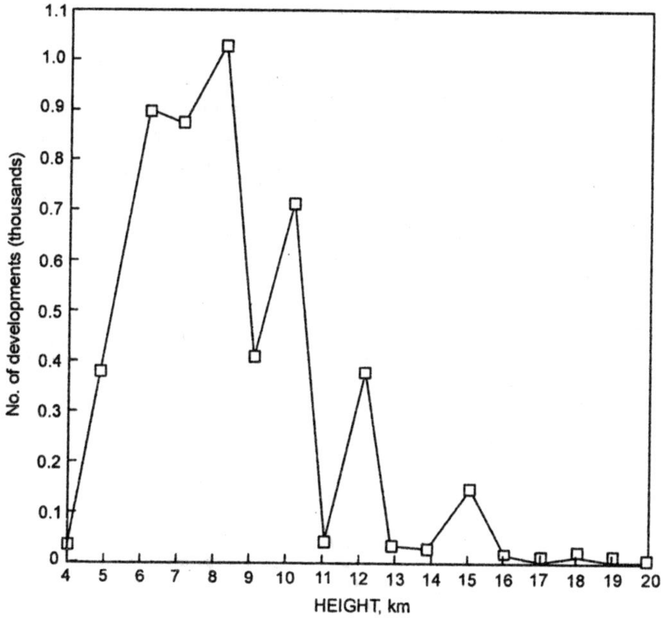

Fig. 7.6: Distribution of Cloud Height for the Entire Pre-monsoon Period During 1980-84

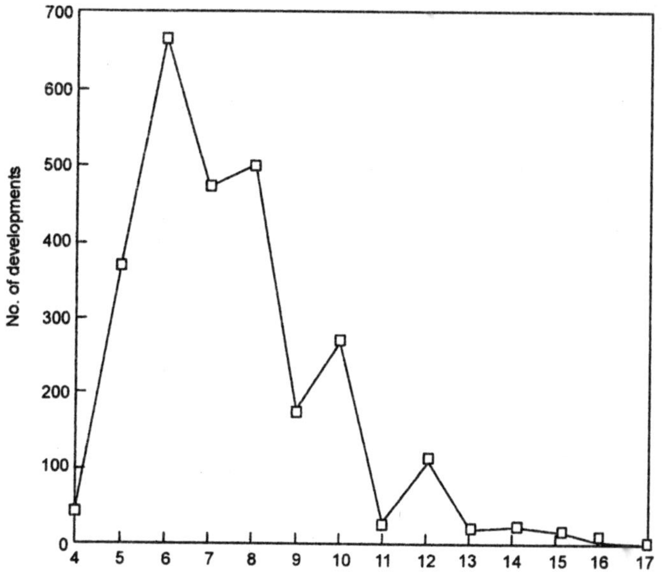

Fig. 7.7: Same as Fig. 7.14, but for the 5-yr period 1985-89

Though DumDum and its surroundings are convectively active during the pre-monsoon period, the decrease in convective developments is supposed to result in decline of precipation.

REFERENCES

Biswas, B. and Gupta, K., (1989), *Mausam,* India, Vol. 40 (2), 169.
Chatterjee, R.N., Prakash, P. and Singh, G., (1995), *Vayu Mandal,* India, Vol. 26, 41.
De, A.C. and Sen, S.N., (1961), *Indian* J. *Meteorol and Geophysics,* Vol. 51 (12).
Koteswaram, P. and De, A.C., (1959), *Indian* J. *Meteorol and Geophysics,* Vol. 10, 275.
Sohoni, V.V., (1931), *Scientific Notes,* India, Vol. 26.

8

RAINWATER HARVESTING: A CASE STUDY OF DELHI

C. Sejwar

INTRODUCTION

The principle of collecting and storing precipitation from a catchment's surface is known as rainwater harvesting. Water harvesting also includes activities aimed at harvesting surface and ground water, preventing of losses through evaporation and seepage as well as other techniques aimed for conservation and efficient utilization of limited water endowment. Rainwater harvesting has been a traditional practice followed by the communities all over the world so as to achieve water security. Today in our country we are witnessing the revival of interest in these age-old techniques.

RAINWATER HARVESTING-CURRENT SCENARIO IN DELHI

The Government of NCT Delhi has realized the importance of rainwater harvesting in Delhi and is therefore undertaking awareness and training programmes for the public. The Government of NCT Delhi has made *water harvesting mandatory* for new buildings which exceed 100 sq.m. The Delhi Government has announced a financial incentive for adopting rainwater harvesting for Resident Welfare Associations (RWAs) and Group Housing Societies. Financial assistance is given upto 50 per cent of the total cost of the project or a maximum of Rs. 50,000/- on satisfactory completion of the project.

In Delhi, Delhi Jal Board (DJB) since 2000 has taken up rainwater harvesting for augmenting ground water storage in NCT of Delhi. A seminar on rainwater harvesting was organized by the Delhi Jal Board (DJB) with the Residents Welfare Associations (RWAs) and Non Government Organizations (NGOs) and other constituent on 30th December, 2002 at India Habitat Centre, New Delhi to further propagate the concept of rainwater harvesting.

The Delhi Jal Board (DJB) has provided rainwater harvesting structures in its 163 building covering an area of about 8 lacs sq. mt., which will harvest about 4.5 lacs cubic metres of rainwater annually for recharging of ground water in Delhi. Work is in progress at another 27 sites of DJB. The Delhi Government has planned scheme for grant of assistance to RWAs/

CGHS for adopting rainwater harvesting structures and provision of Rs. 25 lacs for the annual plan 2002-2003 has been made.

POTENTIAL OF RAINWATER HARVESTING IN DELHI

The growing population and continuously increasing demand for water has put the total water supply system of NCT of Delhi under great strain. The demands of 800 *mgd.** have outstripped the present supply which is about 650 *mgd.* There is, thus a need to augment the water availability to release the pressure on the strained water supply system of Delhi. The ground water resources though being developed extensively hold further scope for augmentation provided the monsoon run off is fully utilized to recharge the depleting aquifers through rainwater harvesting. Artificial recharge of ground water is a process by which the ground water reservoir is augmented at rate exceeding that under natural conditions of replenishments.

The surplus monsoon run off available from Yamuna is 282 m.cm. and runoff is 162 m.cm., thereby surplus runoff available is 444 m.cm. for recharging the groundwater. The water scarcity in Delhi can thus be overcome by harvesting this precious bounty of nature.

RAINWATER HARVESTING

It is a simple economical and eco-friendly method of water conservation. It has been found an ideal solution to recharge the ground water in Delhi. The rainwater that falls on the surface/ rooftop is guided to bore wells or pits or new/old/abandoned wells through small diameter pipes to recharge the underground water which can be used later whenever required.

'W*ater harvesting potential'* is defined as the amount that can be effectively harvested. Water harvesting Potential = Rainfall (mm)* Collection efficiency.

The need for rainwater harvesting in Delhi has been taken up on priority basis due to the following reasons:

(1) Due to rapid urbanization the surface area for infiltration of rainwater into ground has decreased drastically, thereby, diminishing the recharge of ground water.
(2) Over-exploitation of ground water resources has resulted in decline in water levels by 4 to 10m in the last decade alone.
(3) Wells and bore wells are drying up.
(4) Vegetation has also been affected severely.

BENEFITS OF RAINWATER HARVESTING

(1) Quality of ground water improves.
(2) Raises the water level in wells and bore wells that are drying.
(3) Mitigates the effects of drought and achieves drought proofing.
(4) An ideal solution to water problem in areas having inadequate water resources.
(5) Reduces the soil erosion as the surface run off is reduced.
(6) Choking of storm water drains and flooding of roads decreases.
(7) There is saving of energy for lifting ground water due to rise in water table-one metre of rise in water level saves about 0.40 kwh of electricity.

* million gallon daily.

RAINWATER HARVESTING TECHNIQUES

There are two main techniques of rainwater harvesting:

(1) Storage of rainwater on surface for future use.
(2) Recharge to ground water.

The storage of rainwater on surface is a traditional techniques and structures used are underground tanks, ponds, check dams, weirs etc.

Recharge to ground water is relatively a new concept of rainwater harvesting and the structures generally used are:

(1) *Recharge Pits*—Constructed for recharging the shallow aquifers.
(2) *Recharge trenches*—Constructed when the permeable strata is available at shallow depth.
(3) *Recharge shaft*—For recharging the shallow aquifers located below clayey surface.
(4) *Recharge through hand pumps*—The existing hand pumps may be used for recharging the shallow/deep aquifers, if the availability of water is limited.
(5) *Dug well*—Rainwater that is collected on the roof top of the building diverted by drain pipes to a filtration tank, from which it flows into the dug well.
(6) *Recharge well*—Recharge wells are generally constructed for recharging the deep aquifers and water is passed through filter media to avoid choking of recharge wells.
(7) *Roof top rainwater harvesting*—Rainwater falling on roof is collected and injected into the groundwater reservoir.
(8) *Lateral shafts with bore wells*—For recharging the upper as well as deeper aquifers lateral shafts of 1.5 to 2.0 metre width and 10 to30 metre length.
(9) *Spreading techniques*—When permeable strata start from top then this technique is used. Water is spread in streams/nallas by making check dams cement plugs, gabion structures or a percolation pond may be constructed.

Trench with recharge well, shaft with recharge well, percolation tank, check dams, Village ponds, Lakes and depression are also used as structures to recharge ground water.

Table: Major Rainwater Harvesting Projects of Delhi Jal Board

S.N.	*Name of Project*	*Area Harvested (in Sq. m)*	*Qty. of Water (in KL)*	*Techniques Adopted*
1.	Pump House at Deer Park	22100	12155	Recharge Pit with tube well
2.	At Ghitorni STP	39240	21582	Recharge Pits with tube wells
3.	Varunalya Phase-I & II	8800	8250	Recharge Pits with tube wells
4.	Departmental vacant Land agt Burari	9600	52800	Recharge Pits + Trenches + bore wells
5.	At STP Okhla	295800	163000	Bore well and Recharge Trench
6.	Pump House at Greater Kailash	10450	5750	Recharge Pit with tube wells
7.	Office building at Jal Sadan	10000	5500	Recharge Pit with tube wells
8.	Varun Niketan Staff quarters	12500	8125	Recharge Pit with tube wells
9.	Jhandewalan BPS & Reservior No. I & II	6518	4240	Recharge Pits with tube wells

Source: Delhi Jal Board RWH cell Varunlaya Building, Ph-I, Jhandewalan, New Delhi-110005.

Rainwater harvesting had been provided in the Rashtrapati Bhawan, Sharam Shakti Bhawan, Lodhi Garden, I.I.T, J.N.U. in Delhi. The Central Government run Kendriya Vidyalayas (KVs) will soon be equipped with rain water harvesting system. The project will be taken up in all the 849 KVs across India.

COST OF RAINWATER HARVESTING

Typically installing a water harvesting system in a building would cost between Rs. 2000 to Rs. 30,000 for buildings of about 300 sq.m. It is difficult to make an exact estimate of cost because it varies widely depending on the availability of existing structure like wells and tanks which can be modified to be used for water harvesting. The costs would be comparatively less if the system were incorporated during the construction of the building itself. A society with 6 flats may need around Rs. 20,000 to Rs. 30,000 in order to install rainwater-harvesting system.

SUGGESTIONS AND CONCLUSION

The artificial recharge projects would be taken up all over Delhi to check the decline in water levels. In case of continuous decline of water levels the pumping should be regulated. The practice of rainwater harvesting should be adopted by all industries and the farming sector to avoid twin problems of flooding due to bad drainage and water shortage. Especially over-bridge and highways which cover a large area, can be used as harvesting surfaces and the water can be collected in ponds for further use.

In the end, it is concluded that Rain Water Harvesting (RWH) is a viable technology and can become an integral habit.

REFERENCES

Aggarwal, A., (1997), Dying Wisdom, *Centre for Science and Environment (CSE).*

Jain, M.K., (2001), Yojana-Rainwater Harvesting, Aims Techniques.

Mishra, A., (2001), *Tehera Pani Nirmal Parti Bhoomi Samachar*, January-March.

Proceeding of the seminar conducted by Delhi Jal Board (DJB) on 30th December 2002 at India Habitat Centre, New Delhi.

Sen, P.V. and Sen, P.K., (2001), Proceeding of the Workshop on 'Domestic' Roof Water Harvesting Sponsored by European Union and Rajeev Gandhi National Drinking Water Mission, I.I.T. Delhi.

SECTION III

LANDUSE AND LANDCOVERS

9

ECOSYSTEM MANAGEMENT AND DYNAMICS OF LANDUSE/LAND COVER IN DAKSHINA KANNADA DISTRICT (1970-1971—1999-2000)

AFSARI JAN AND M.R. VIJAINDRA

INTRODUCTION

Ecosystem management is a process that integrates biological, social, and economic factors into a comprehensive strategy aimed at protecting and enhancing sustainability, diversity, and productivity of our natural resources. Ecosystem management means looking at the big picture, and working closely with other land mangers, both public and private. It means addressing the long term consequences of today's decisions, and thinking of various resources as interrelating parts of systems rather than as individual components to be managed separately. It means awareness of many scales of effect, from local and national to international and even global. The tough choices would not go away. However, a fundamental principle of ecosystem management holds that decisions must be based on the best information which can be provided with sustainability ecosystem, and which will require everyone's support and participation to make it all work. Historians, archaeologists, and cultural anthropologists routinely gather and interpret data about past environment and the ways humans have altered or been affected by them. Without an understanding of human effects on the environment through time and the long – term natural cycles of change, decisions on how to manage an ecosystem can be subject to error. Archaeological data can tell us which plants and animals were present at a particular place and time. Such data are available for North American cultures for the last 12,000 years. Resource managers rely on this information when planning future plant and animal conditions within a given ecosystem.

Other tools archaeologists use to reconstruct past climates are tree-ring studies, soil analysis and geomorphologic analysis. Through these studies, information is gleaned on vegetation composition, soil conditions, and drought cycles. Such data are helpful in distinguishing natural from human-induced environmental changes. Historical records, such as old photographs, survey records and surveyors' note, and oral histories are invaluable for understanding more recent impacts of humans on the environment. Surveyor's notes record vegetation and animal population just prior to the period of rapid westward expansion, providing a detailed "snapshot" in time. Similarly, comparison of photographs taken from the same vantage point decade's apart document landscape changes.

Though ecosystem which man enters and adjusts is quite interesting from the Sahara's with 50°C temperature with the African tribes to the Chukchi's of Tundra living in – 70°C temperature. Man has adapted himself, and not just that he has grown, gained knowledge, skills and designed lifestyles to suit to those conditions which carve out the landuse patterns around him. This gives an insight as to how man adapts to his surrounding environment making adjustment around himself.

This case study tries to briefly touch on the landuse changes and ecological management, in Dakshina Kannada district of Karnataka.

STUDY AREA

The coastal district of Karnataka-Dakshina Kannada and Udipi has been taken into consideration for the study. Primarily both the districts were one but in 1997 they were divided into two. Therefore, in this case study the combined area of both the districts are taken in to account. The study area lies roughly between 74°35′ to 75°50′ N. Latitude and 12°27′ to 13°58′ E Longitude.

The taluks of Sullia, Puttur, Buntwal, Mangalore, Belthangdi, Karkal, Udipi and Coondapur form the study area. The total area of the district is 8441 square kilometres with 3005897 population giving a density of 351 persons per square kilometre. The location of the study areas is shown in Figure 9.1.

CLIMATE

The entire region experiences coastal climatic conditions coupled with the monsoonal influxes. The temperature ranges between 27° to 34°C. The rainfall ranges between 200 to 400 cms. This region has high humidities and an oppressive weather in the hot season. There are considerable climatic variations between the areas near to Western Ghats and those near to the Arabian Sea and are those that actually on the coast. Hence, the season can be divided in to the following categories:

March to May: Hot Summers
June to August: Heavy rains
September to October: Pleasant season
November to December: Cool Season
January to February: Cool and Crisp Season

PHYSIOGRAPHY

The districts comprise a long strip of territory running from south to southeast and are broader towards south. The lands belonging to the Kerela state in fact bound the taluks namely Sulya, Puttur, and Buntwal on the west. The district, which is about 177 km in length, has a coastline of about 140 km. The breadth varies between 10 km at the narrowest northern most tip and about 80 km at the broadest portion. The landuse forms a broken low plateau spreading out from the foot of Western Ghats towards the numerous valleys containing rich alluvial

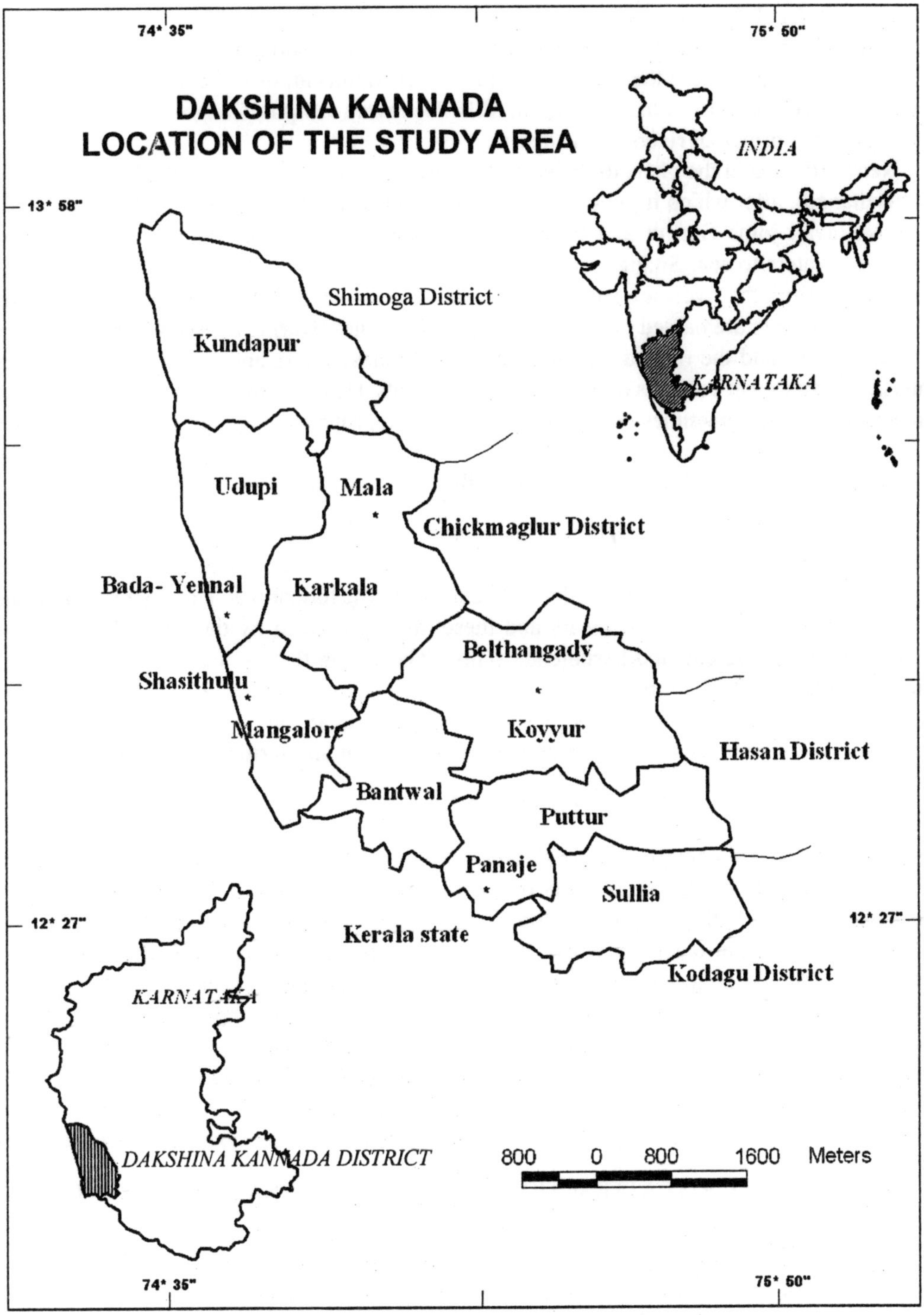

Fig. 9.1: Dakshina Kannada Location of the Study Area

soils. The peaks of the scarp edges vary in altitude from 910 metres to 1830 metres above mean sea level. The most interesting of the mountains here is the Kudremukh group, which consist of three peaks. Apart these there are numerous detached hills all over the district. The coastal strip is relatively free from these detached hills. Netravati, Gurpur, Gangoli, Sitanadi and the Swarna are the principal rivers of the district.

Netravati rises in the ghats to the east of Kudremukh and flows down the Bangadi valley past Beltangdi, after which it joints Kumaradhara river near Uppinangady. At Mangalore, the river Netravati joins Gurpur and forms large estuary through which both discharges their combined waters to sea. Sitanadi and Swarnanadi join near Barkur to form a large sheet of backwaters. Both the rivers are navigable for a distance of 18 kilometres from the coast. Most of the minor rivers are named on the corresponding villages. During the monsoon, because of the heavy rains and the rugged nature of the terrain, these rivers and streams overflow their banks and carry great volumes of water in rapid currents. Heavy rains and the resultant flooded streams and rivers sometimes cause damage to life and property including the standing crops.

SOILS

Geologically, with the exception of the coastal strip, the district contains rock formation belonging to the earliest period of the earth's history, namely the Archean epoch. The coastal region contains recent and sub recent deposits and the latertic formations. The older metamorphic rocks constitute mainly the Dharwars and these are represented by the banded ferruginous quartzites and by the talc and hornblende schist. The soil in the district is of a lateritic type characterized by a high iron and aluminium content. This variety usually occurs in heavy rainfall zone and is suitable for crops such as paddy, sugarcane, arecanut, cardamom and plantains. The numerous valleys and ravines, however, contain fertile alluvial soil.

FLORA AND FAUNA

The climatic and soil conditions have all along promoted the spontaneous growth of a variety of plants and trees in almost all parts of the district. The slopes of Western Ghats are, in particular, clothed with dense vegetation. In the coastal belt the tree growth is sparse. The chief timber trees are teak, rosewood, benteak, casuarinas, cashew etc. The abundant tree growth provides shelter to a variety of wildlife. Bison, sambar, spotted deer, wild bear, monkeys, rabbits, foxes, porcupines, mongoose, squirrels etc., are amongst the important types of animals found in this district. Amongst the reptiles cobra and the harmless green whip snake are to be found in all parts of the district. The rivers, streams and the sea is bound with variety of fishes.

Nature has blessed this district with an abundance of rainfall, which seldom fails to materialize. As a result, famine and scarcity conditions are almost unheard of in this district.

MAJOR TOURIST ATTRACTION IN DAKSHINA KANNADA AND UDUPI

Gomateshwara

Venoor is about 19 km from Belthangady and about 69 km from Mangalore. The chief objects of interest here are the colossal monolithic status of Gommata of 38 feet in height and a few basadis. The Gommata statue was set up in 1604 AD by Veera Thimanna Aiila IV.

Udupi Sri Krishna Temple

Udupi is a divine shrine in the coastal region of Western Ghats. Situated about 60 kms from Mangalore, it is a fairly large and upcoming town that was sanctum of Madhwacharya, the great Sanskrit Philosopher. The famous temple here, has a fascinating idol of Lord Krishna that is richly adonred with jewels. It is the centre for education, social and religious activities in the district of South Kanara.

Kateel Durga Parameshwari Temple

Situated on the banks of River Nandini, amidst beautiful surroundings, it is the spiritual energy centre of South Kanara district. Goddess Durga is the loving and caring mother of the people of South Kanara. Surrounded by lush green fields and beautiful hills, Kateel has become a famous pilgrimage and a tourist centre over the years.

Dharmasthala Manjunatha Temple

Situated on the banks of river Nethravathi and surrounded by the lush green forests and hills of Malnad, Dharmasthala is known for its beautiful surroundings and Sri Manjunathaswamy Temple. In Dharmasthala, charity is a way of life, extend to all without discrimination. Sri Dharmasthala is one of the sacred places in south India and is a famous pilgrim centre. It is situated in the village of Dharmasthala, in Belthangady Taluk of South Kanara district about 75 from Mangalore.

Mangalore

Mangalore is a long stretch of beautiful green land at the southwestern part of India peninsula. The word 'Mangalore' means auspicious place. 'Mangala' (Sanskrit) means happiness auspicious, 'Uuru' the end suffix means 'place'. 'Mangalore' is the Portuguese corruption of the word 'Mangaluru' or 'Mangalapura'. But the natives identify this place as 'Kudla' or Kodial meaning the joining of two rivers. 'Kuudu' means to join and 'ala' refers for water/ river. Mangalore is well connected with the country air, rail and road. The stretch between Mangalore and Udupi can be covered in an hour by bus. There is also train facility to Udupi. The New Mangalore Harbour with the iron ore projects, and iron pellet processing plant can be seen at New Harbour. For sight seeing and picnic one can go to the picturesque Ullal beach just 10 kms from the Heart of Mangalore City. 'Summer Sands' a beach resorts with beach side restaurant and cottage in native styles attracts tourist.

Maravanthe Beach

This scenic beauty is situated 105 km. away from Mangalore. This is a unique place where the Arabian Sea and the river is separated by a narrow road.

Panambur Beach

This beach is situated on the Mangalore-Udupi highway (NH-17) to the left around 7 kms from the city. On the way to this beach one can see MCF (Mangalore Chemical Fertilizer) and

Kudremukh Iron Ore Company. The Panambur beach is a popular Sunday outing spot among the locals. This beach is very near to the New Mangalore port, which is the harbor for many ships. Usually granites and grains are exported from this harbor. This is a beautiful clean beach attracting large number of visitors and tourist during summer vacation and other holidays. One can also view splendid ships sailing among the ups and down of the rough Arabian sea. Chat and snack stalls are available near the beach. One can see the shoreline stretching of all the way to Surathkal from here. The lighthouse of the Surathkal beach can also been seen from this beach.

Surathkal Beach

This beach is also situated on the Mangalore-Udupi highway (NH 17) to the left around 20 km from the city. This beach is opposite to the Karnataka Regional Engineering College (KREC). The main attraction of this beach is the Surathkal Light House. In the evening one can find many engineering student gather here for a small walk and for some games. It is a pleasant place for studies, a very peaceful and a calm beach.

Thaneerbhavi Beach

This beach is also situated on the Mangalore-Udupi highway (NH 17) to the left around 6 km from the city. Kudremukh Iron Ore Company comes on the way to this beach, from the sidewalk one can have a look at the Iron Ore Company. The special attraction of this beach is the boating facility. Boating is open to public only during summer and during public holidays. You can take a panoramic boat ride through the backwaters of the Arabian Sea.

LANDUSE AND LANDCOVER

The study of landuse/landcover from the 1971 to 1999 has been done using secondary data collected form various sources of government organizations. Table 9.1 shows the percentage area covered by each landuse to total area with a decadal cascade.

Table 9.1: Landuse/Land cover in Dakshina Kannada (Figures indicate percentage)

Landuse/Landcover	*1971*	*1981*	*1991*	*1999*
Forest area	27.65	27.15	27.2	27.31
Net Sown Area	22.83	23.08	26.23	27.45
Permanent Pastures	4.32	3.97	3.81	8.2
Land under Miscellaneous and Tree Groves	11.91	11.35	10.85	3.68
Land under non agricultural use	8.94	9.98	10.27	11.25
Cultivable	9.52	9.81	8.46	10.17
Current fallow	4.29	2.67	2.00	1.66
Barren and Uncultivable land	8.17	8.71	8.65	8.53
Other Fallow	2.37	3.28	2.53	1.75
TOTAL	100.00	100.00	100.00	100.00

The forest area of the study region has remained static preserving its spatial extent with its varied vegetation and its sanctuary for the wildlife. There has been overall increase in the net sown area from 22.83 per cent to 27.45 per cent which is very normal as the population has also undergone change. There has been 4 per cent increase in the land under non-agricultural use from 8.94 per cent to 11.25 per cent. There has been very little change in cultivable wasteland and barren land. The following table shows the ratio between the land covers between the study region and Karnataka State.

Table 9.2: Percentage of Landuse/Landcover in Dakshina Kannada and Karnataka State

Landuse/Landcover	*Karnataka*	*Dakshina Kannada*
Forest area	16.06	27.31
Net Sown Area	56.59	27.45
Permanent Pastures	5.12	8.2
Land under Miscellaneous and Tree Groves	1.63	3.68
Land under non agricultural use	6.86	11.25
Cultivable	2.2	10.17
Current fallow	6.64	1.66
Barren and Uncultivable land	4.35	8.53
Other Fallow	0.54	1.75
TOTAL	100.00	100.00

When we compare the percentage values of the state and district with respect to the total area of each unit the study area is much well balanced compared to the state. But the agricultural land in the district is a little over than one fourth area of the district which is a negative per cent from an agricultural point of view.

ECOSYSTEM MANAGEMENT STRATEGY FOR THE STUDY AREA

There is substantial user demand for natural resources from many different levels ranging from village councils to state and central governmental agencies as well as NGO's. India now has evolved a vigorous system of decentralisation where more and more powers of management of natural resources are being devolved to village councils. These councils are interested in natural resources and ecosystem goods and services assessment. The governments are encouraging this. The proposed ecosystem assessments would help develop the relevant experience for better management and safeguarding of the ecosystem. The anticipated user groups include central and state governments, local organization such as youth club, women's club, farmers' union, fishermen's co-operative, forest labourers' society and local educational institutions.

ECOSYSTEM SERVICES

Ecosystem services highlighted here include the use of streams for irrigation, free grazing of cattle, pollination, recreation, cultural and spiritual satisfaction. Ills of an Ecosystem include pollutants, exotic invasive weeds, and vectors of diseases. Ecosystem disservice's of interest include: crop raiding by wild animals. Activities that will be examined include use of pollution of water upstream, sale of agrochemicals to farmers, operation of a fish cold storage, eco-tourism, pilgrimage, poaching, and control over reserved forest or agricultural development programmes. Agriculture, coastal fisheries and collection of non-timber forest produce are the mainstay of the economy of the study sites. As a result ecosystem goods and services, bads and disservice's play a significant role in the lives of the people. Ecosystem goods highlighted in this assessment include: fish, NTFPs, medicinal herbs, leaf litter, shells and sand. This work is in 11 stages, which is being carried out by centre for ecological sciences in various other areas also.

MONITORING

Scientific information gathered through inventory, monitoring, and assessment forms the basis of ecosystem management. As one might imagine, tracking and evaluating the effects of multiple actions on something as complex, interactive, and dynamic as an ecosystem is no simple matter. To define even the scale of a particular ecosystem is fraught with controversy—should we look at continents, geographic region, watersheds, vegetative communities, small study tracts, or soil microbe communities? Scientist have come to realize the need to look at all of them. But how can we spend our time and money collecting appropriate data? The invasion of undesirable are weed species when grasslands are overgrazed and change in aquatic plant and animal communities in response to population. More commonly, however, the signals of ecosystem deterioration are not so easy to read. Choosing the right data to monitor is crucial to whether or not we can detect ecosystem instability. Once the most useful data has been identified, the next challenge is how to collect and store that information. Data should be collected in a rigorous and consistent manner. Data collection procedures are beginning to be standardized across administrative boundaries, but there is still a long way to go on broad scale efforts, especially international ones.

Some current techniques for collecting monitoring data include laboratory analysis of water, soil, and air samples for contaminants and various life-forms, inventories of plant and animal species; genetic studies; sensors in smokestacks; and satellite sensors. To assess how ecosystems have changed over time, and to track natural cycles that may span centuries, scientists tap historic data, such as old photographs and diaries, and data from archaeological sites. Fossil pollen, present in soil, reveals ancient plant community composition, and even ancient garbage dumps provide surprising clues about the plants and animals that once lived in an area.

Interpreting the collected data is, perhaps, the most difficult aspect of monitoring. Information from a variety of sources must be synthesized and integrated to paint a complete picture. And, since environmental science is a relatively new endeavour, the implications of certain data are not clear-cut. Researchers are constantly devising new and better technologies for tracking the condition of environment. (In fact, this is a fast-growing industry, one with abundant career opportunities.) Ideally, monitoring and assessment give the resource manager or community decision-maker constant feedback about the effects of management practices and their relationship to long-term sustainability. This permits adaptive management, or ongoing adjustment, so that "environmental train wrecks" can be avoided.

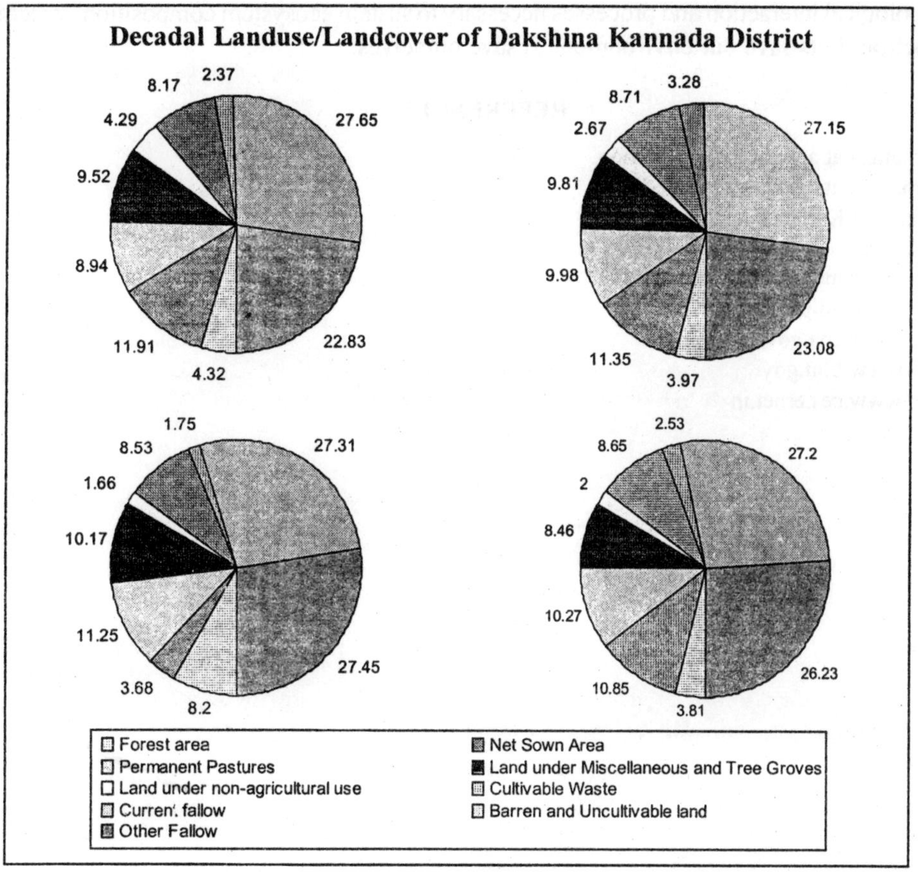

Fig. 9.2

RESULT

There is a very bright side to the whole study, whenever any one talks about our ecosystem and its balance we always talk as to how we are successfully choking ourselves and jumping in to the black hole but, here an attempt has been made to say that all is not lost. The study here has a forest cover, which is 27 per cent. This is a highly remarkable bit of information, which we can say is the outcome of not the government but the people. Finally, worldwide people now realize the value of ecosystem and are joining hands of reconstructing what they so miserably have destroyed. But, the people of Dakshina Kannada had known its value all along.

The ecosystem of this region is being given more importance by the local non-governmental organizations there and they are working to make it even more safe and sound for all types of life. Its high time the country realizes what all this aping the west had done to us.

CONCLUSION

Ecosystem management is management driven by explicit goals, executed by policies, protocols, and practices, and made adaptable by monitoring and research based on our best understanding of the ecological interaction and processes necessary to sustain ecosystem composition, structure, and function. Lets save our environment to save ourselves.

REFERENCES

1. Karnataka at a glance, 1994 edition.
2. Karnataka at a glance, 2000 edition.
3. Census of Karnataka, 2001 edition.
4. Websites
   (a) www.mangalore.com
   (b) www.udipi.com
   (c) www.epa.gov
   (d) www.blm.gov
   (e) www.ces.ernet.in

10

REMOTE SENSING AND GEO-INFORMATION SYSTEM IN SUSTAINABLE DEVELOPMENT OF LAND RESOURCES:
A Case Study of Yacharam Watershed in Ibrahimpatan Block, District Rangareddy, Andhra Pradesh

DILIP KUMAR AND DR. SAROJ K. PAL

INTRODUCTION

Applications of remote sensing and GIS techniques have gained importance during last two decades for deciding many issues related to natural resource mapping, monitoring and planning. Remote sensing technology because of its synoptic view and repetitively has emerged as one of the most powerful tool for the study of natural resources. The high speed, accuracy and the temporal nature of the remote sensing data make it more useful for this study over conventional techniques.

Watershed is a natural hydrologic dynamic unit with a certain extent of homogeneity and uniformity and is a limited convenient clearly defined unit. The concept of a watershed as a planning unit for development of land and water resources has gained importance since 1974 when Ministry of Agriculture, Government of India initiated various developmental programmes like Drought Prone Area Programme (DPAP), Desert Development Programme (DDP), Hill Area Developmental Programme (HADP) etc. Therefore, it is necessary to delineate watershed boundaries at various levels of hierarchy to identify developmental activities under various schemes in each watershed. Drainage network helps in delineation of watershed and for suggesting various water harvesting structures and soil conservation measures.

Sustainable development aims at maintaining equilibrium between the human needs and economic development within the parameters of environment conservation through efficient use of natural resources to ensure trade-off between desired production and consumption levels (Sahai, 1993). The well-known Brundtland Commission defined sustainability as a "Development that meets the needs of present without compromising the ability of future generations to meet their own needs. In essences, sustainable development is a process of changing in which the exploitation of resources, the direction of investment, the orientation of technological development and institutional changes, all are in harmony". Sustainable development of resources based on maintaining the fragile balance between productivity functions and conservation practices through monitoring and identification of problem areas, agricultural practice, crop rotation, use of bio-fertilizer, energy efficient farming methods and reclamation of under utilized land. Sustainable development requires a holistic approach towards harnessing natural resources after taking into account the precarious environmental condition. (NRSA, 1995).

The exploitation of natural resources beyond their regenerative capacity has led to global environmental degradation, stretching the life sustaining mechanism beyond their recuperative limits. In densely populated India, human beings occupy every inch of space within the physical environment for subsistence and survival. Also, rapid growing population creates heavy demand on natural resources and the production system. Thus, increasing emphasis is being laid on scientific management of natural resources to ensure their optimal utilization, keeping in view conservation, environmental and socio-economic need for conservation at micro-level. It needs understanding of the mutual interdependence of land and water resources, characterization and identification of the constraints and understanding of ecological problems at the micro-level.

Therefore, one should look for the optimum use solutions regarding sustainable development of resources. The approach method for studying land resources at watershed level through the use of remote sensing and GIS helps arriving at locale specific prescriptions to achieve sustainable development of natural resources of the region.

OBJECTIVES

The ultimate goal of the study is to generate site-specific action plan by integrating natural resources information generated from satellite data in conjunction with collateral data.

STUDY AREA

Yacharam watershed is part of Ibrahimpatan block in Rangareddy district of Andhra Pradesh as shown in Fig. 10.1. The geographical extent of the watershed lies between 78° 35' 00" to 78° 45' 00" east longitude and 17° 00' 00" to 17° 5' 00" north latitude. The Yacharam watershed has an areal extent of approx. 134 sq kilometres. The study area is bounded on the east by Nalgonda District and the south by Mahaboobnagar District. It comprises Chaudharpalli, Namdiwanparli, Medpalli-Nakkerta and Chintapatla.

The survey area appears to be an erosional landscape in an advanced stage of pediplanation. It has gently undulating topography; erosion and deposition are the two most important physiographic process. Hills are remnants of natural dynamic process of denudation and weathering aided by fluvial action, which occur as massive hills, tors and domes scattered throughout the project area rising the height of 600 to 700 metre a bove mean sea level. Most of the project area constitutes a pediplain carved out of Achaean rocks.

The climate of the study area is characterized by a hot summer and a generally dry except during south-west monsoon season (June to September). From February, both the day and night temperatures begin to increase steadily. May is the hottest month with mean maximum daily temperature of 40°C. With the onset of the south-west monsoon into the early in June there is appreciable drop in temperature, and weather becomes more pleasant. There is a considerable drop in temperature in the beginning of November with the mean daily temperature at 29°C and the mean daily minimum temperature at 13°C.

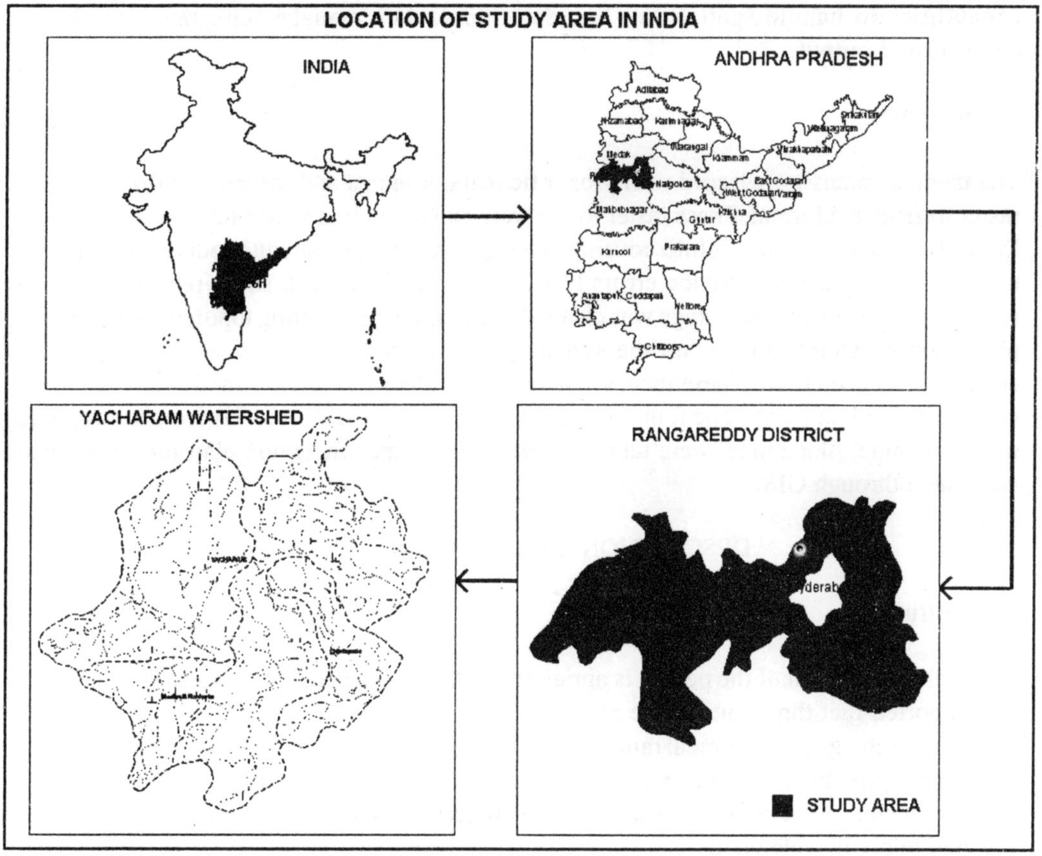

Fig. 10.1: Location of Study Area

DATA SOURCES AND METHODOLOGY

Data Source

The thematic maps were collected from the ISD division, NRSA, Dept. of space which were prepared by A.P. State Remote Sensing Application Centre, Hyderabad from satellite data IRS-IB LISS-II and survey of India toposheet No. 56K/12 and 56L/9.

Software used: Workstation based (IBM AIX 3.2) ARC/INFO 7.03 GIS Software developed by ESRI, USA and Workstation Based Erdas-Imagine 8.3 developed Erdas, USA in SGI, IRIS 4.0 operating platform.

For the case study, GIS mainly served as an instrument to store and analyse a wide range of natural resources data. The main purpose of using GIS was to prepare the action plan for utilizing the land resources in a conservative, economic and productive manner. GIS is a computer-based system designed to accept large volume, manipulate, analyse and display various data according to the user-defined specifications. It is capable to handle and analysis data that are referred to geographic location. A GIS provide four type of capabilities to handle spatial data i.e. data input, data management, manipulation and analysis and output.

Methodology

The thematic maps were traced with proper tic mark, scanned and loaded into the computer in raster form in. tiff format. These raster forms were converted into vector form through ERDAS-IMAGINE package, and ArcInfo coverage was generated. The Acredit module of ArcInfo was used to identify and remove the errors like overshoot, undershoot, sliver polygons etc. After removing the errors, the coverage was cleaned and build for generating topology. The polyconic projection in ArcInfo to convert the system coordinate value into real world coordinates of coverage has been used. These thematic maps were integrated one by one for development of composite land unit map. The four thematic layers, landuse/landcover, Hydro-geomorphology map, Soil map, Slope map, were taken to generate site-specific action plan for the Yacharam watershed through GIS.

DESCRIPTION OF NATURAL RESOURCES

Agriculture

The main occupation of the people is agriculture, which is dependent on vagaries of monsoon. It is reported that three out of five monsoon years are subnormal with deficit ranging from 25 – 35 percentage below normal rainfall. The rainfall is also erratic and not uniformly distributed to suit crop growth period and crop water requirement. Long dry spells lasting 3 to 4 weeks are not uncommon. As a consequence crops are subjected to severe moisture stress resulting in low and unpredictable yields.

Geology

The area is underlain by granite rocks of Achaean age, gray and pink in colour and medium to coarse in texture. These rocks are exposed as high-elongated hill ranges, linear ridges, tors and domes in many places. The rocks have prominent joints trending northeast to southwest with a dip towards northwest. The granites are intruded by dolerite dykes having two major trends: one from north to south; and the other from west to east. The material of local alluvium occurring along the streams is also derived from granite.

Drainage

Yeliminedu vagu and pedda vagu and their numerous tributaries, which are seasonal in nature, drain the area. The drainage intensity in the hilly areas and upper slopes is high and moderate in the lower slope. The drainage pattern is dendritic to sub dendritic as shown in Fig 10.2.

HYDRO-GEOMORPHOLOGY

The study area can be described as a vast pediplain dotted with hills and intruded with dolerite dykes. The geomorphological forms identified are given below and shown in Figure 10.3 and Table 10.1.

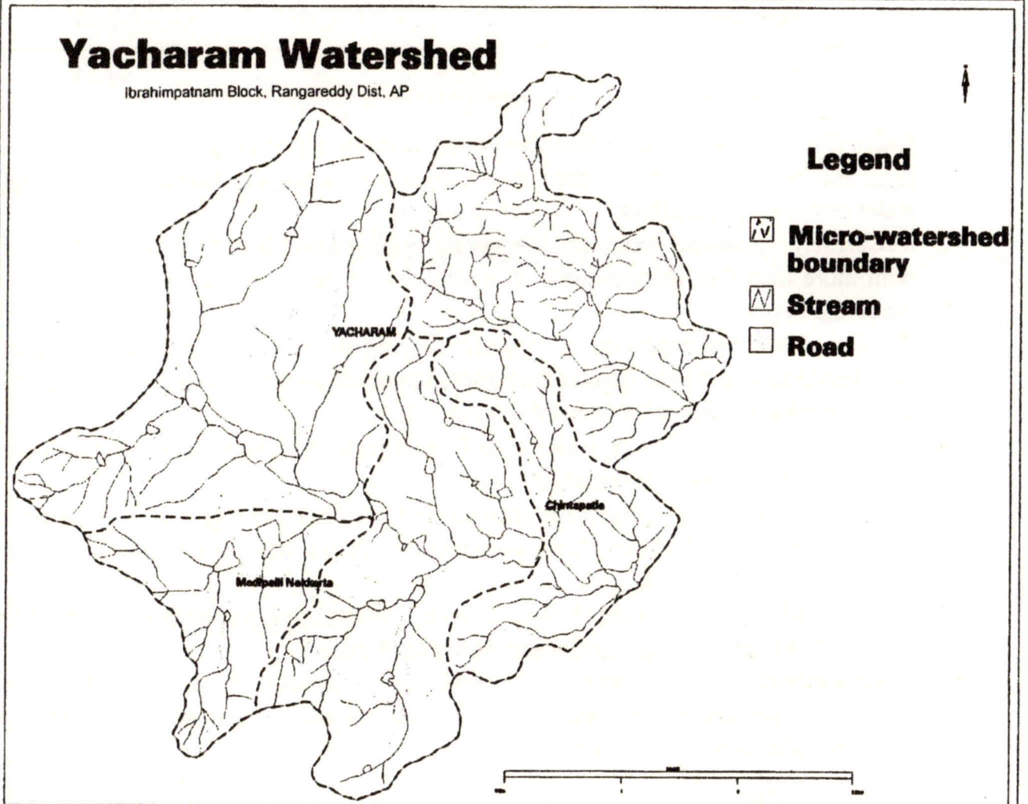

Fig. 10.2

Table 10.1: Showing Geomorphic Units and their Area

Sl. No.	*Geomorphic Unit*	*Area (in Sq. Kilometre)*
1.	Shallow weathered pediplain	96.27
2.	Tor/tor complex (pg)	1.18
3.	Rocky pediments (clg)	1.57
4.	Residual hills (clg)	8.48
5.	Moderately weathered pediplain	8.31
6.	Rocky pediment (pg)	2.31
7.	Tor/.tor complex (clg)	1.06
8.	Residual hills (pg)	2.19
9.	Valley fill	9.33
10.	Waterbodies	3.44
	TOTAL	134.14

(1) *Valley fill*: it is an unconsolidated sediment comprising of cobbles, pebbles, gravel, sand and silt deposits by stream river normally in a narrow fluvial valley, ground water potential is excellent.

(2) *Moderately weathered pediplain*: it is gently sloping smooth surface of granitic gneiss with more than five metres depth of weathered material. Ground water potential is moderate to good.

(3) *Shallow weathered pediplain*: it is gently sloping smooth surface of granitic gneiss with less than five metres depth of weathered material generally covered with red soil. The ground water potential is poor to moderate.

(4) *Pediment*: it is gently sloping rock cut surface of granites and gneisses with thin veneer of detritus. Ground water potential is poor.

(5) *Tor/tor complex*: (pg = peninsular granites and gneisses and Clg = closepet granite): these are a group of spheroidally weathered boulders with isolated rock outcrops. Ground water potential is poor.

(6) *Residual hills*: it is an isolated low relief relict hill occupying considerably small area formed due to erosion. Ground water potential is poor.

(7) *Dyke/dyke ridge*: these are vertical to steeply dipping dolerite dykes intruded in granite basement. They are massive, highly, jointed, narrow linear ridges or heap of boulders. Ground water potential is poor.

Soil

Soil is the one of the basic component of the agricultural production system. Agriculture is the main occupation in the area. Hence the potentials and limitation of soil for agriculture and various other uses and its response to fertilizers and irrigation are determined by the soil characteristics, which to a large extent are expressed by texanomic classes. The soil in the area basically are red soil and fine, loamy typic, lithic. Depth varies from shallow to deep. The type

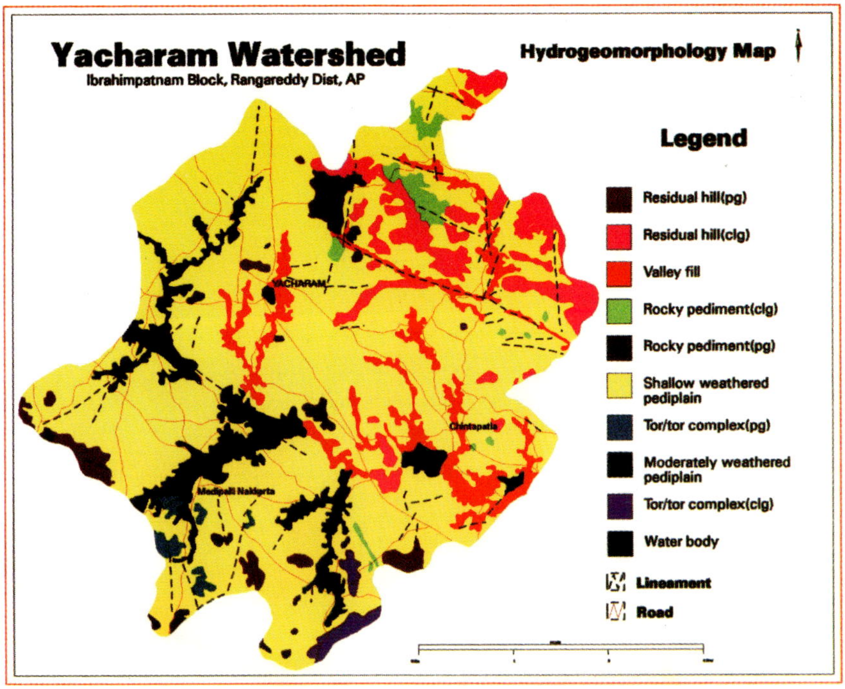

Fig 10.3: Hydro-Geomorphology Map

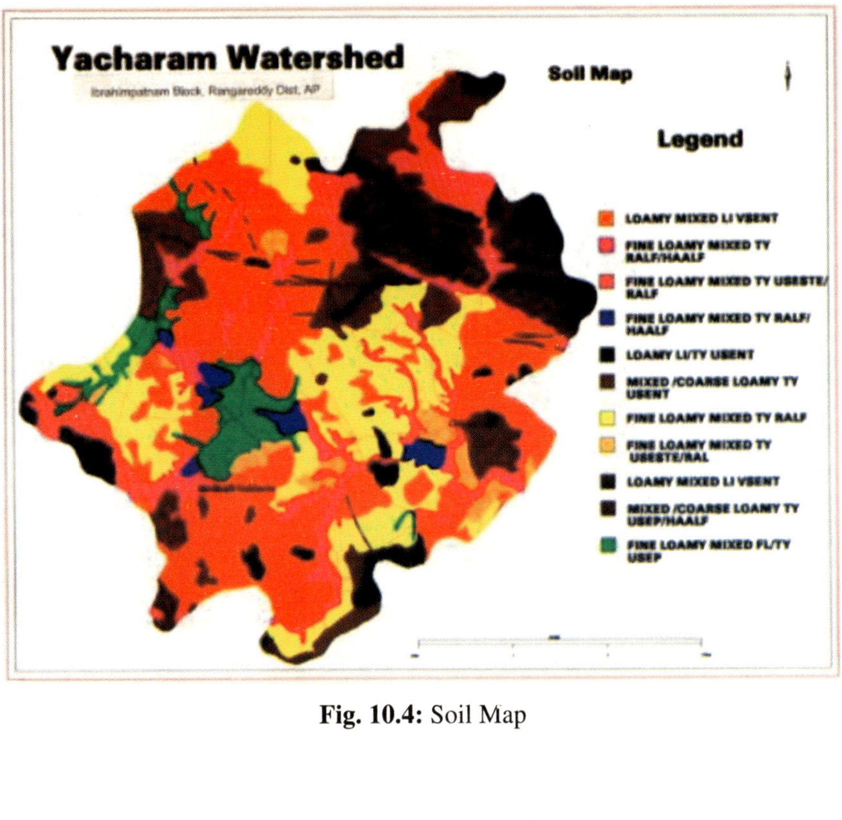

Fig. 10.4: Soil Map

of soil with its taxonomic classification and the area occupied by each type is shown in Table 10.2 and Fig. 10.4.

Table 10.2: Showing Soil Classes and their Area

Sl. No.	*Soil Class*	*Area in Km²*
1.	Loamy skeletal/mixed, lithic/typic ustorthents	12.38
2.	Loamy skeletal/mixed, typic ustroepts/haplustalfs	9.43
3.	Loamy skeletal/coarse mixed, typic ustorthents	13.51
4.	Loamy skeletal/mixed, typic ustroepts/haplustalfs (rangapur soil series)	7.07
5.	Loamy skeletal/fine loamy, mixed typic rhodustalfs/haplustalfs	38.12
6.	Finel loamy/loamy skeletal, mixed typic rhodustalfs	24.94
7.	Fine loamy/loamy skeletal, mixed typic ustorpets/rhodustalfs	3.47
8.	Coarse to fine loamy, mixed typic ustorpepts	14.13
9.	Fine loamy, mixed fluventic typic ustorpepts	6.54
10.	Loamy skeletal, mixed typic ustorthents	1.10
11.	Waterbodies	3.44
	TOTAL	134.14

Landuse/landcover

The brief description of landuse/cover categories along with the areal extend is given in Table 10.3 and Fig 10.5.

Table 10.3: Showing Landuse/Cover Classes and their Area

Sl. No.	*Landuse/Cover*	*Area (in Sq. Kilometre)*
1.	Built-Up Area	0.45
2.	Kharif Unirrigated	92.92
3.	Double Crop	9.34
4.	Fallow Land	0.39
5.	Plantation	0.09
6.	Degraded/Under-Utilized Forest	10.86
7.	Land With Or Without Scrub	13.62
8.	Barren/Stony Waste/Sheet Rock Area	1.08
9.	Mining Area	0.88
10.	Salt Affected Area	1.07
11.	Water Tank	2.5
12.	Dry Water Tank	0.94
	TOTAL	134.14

Built-up land: area of human habitation, which covers building, transport, communications, utilities etc.

Agricultural land: the land primarily used for agriculture purposes and production of food, horticulture and other commercial crop. It also contains land under fallow, plantation, etc.

Wasteland

Cultivable wastelands are described as degraded land, which can be brought under vegetative cover with reasonable effort. It also contains land with or without scrubs, barren rocky/stony waste/sheet rock area.

Water bodies: these are surface water in the form of ponds lakes and reservoirs.

Others: all other uses that do not come under above categories are grouped under other category. Like mining area, mainly quarries and limestone mines.

Slope

The slope map depicts steepness of the land and its tells about the terrain characteristics. Slope information is useful in deciding the soil conservation, management practices. For Yacharam watershed undulating terrain exists mostly in the study area. Slope ranging from gently sloping to high slope category as shown in Table 10.4 and Map 10.6.

Table 10.4: Showing Slope Classes and their Area

Sl. No.	*Slope Class*	*Slope in %*	*Area (in Sq. Kilometre)*
1.	Nearly level	0 to 1	22.32
2.	Very gentle	1 to 3	80.42
3.	Gentle	3 to 5	16.31
4.	Moderate	5 to 10	9.42
5.	Moderate steep	10 to 15	3.77
6.	Steep	15 to 35	1.9

RESULT AND ACTION

The action plan for the land resources development is given in action plan Table 10.5. As per the suitability of the resources such as soil, water, slope, 21 hectares of the area is suggested for double cropping. Agro horticulture has been proposed in the area where water availability in slightly less but other resources is good. Intercropping is suggested for approx 7 hectares where soil is good, slope is 3-5 per cent and water availability is moderate. Stabilization of mining area has been suggested for mining area.

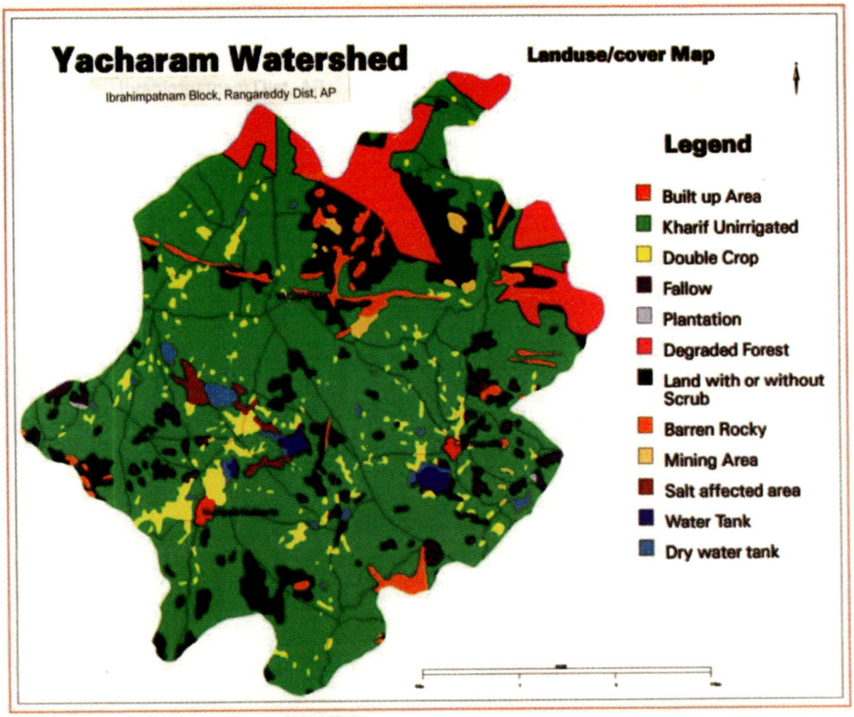

Fig. 10.5: Landuse/Cover Map

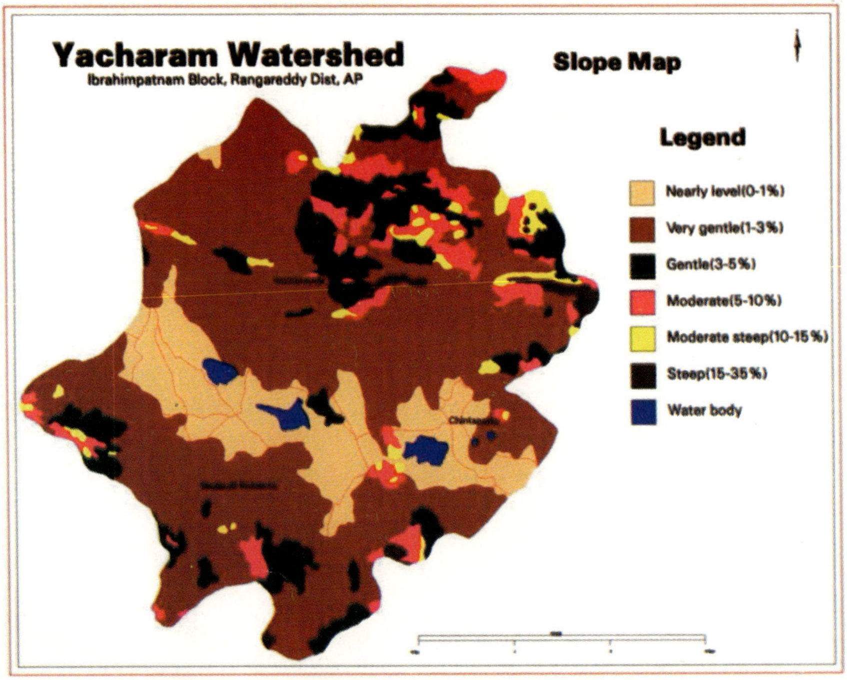

Fig 10.6: Slope Map

Table 10.5: Decision Rule for Action Plan

Sl. No.	*Existing Landuse/Cover*	*Hydro-Geomorphology*	*Slope*	*Soil*	*Recommended Landuse System*
1.	Kharif Unirrigated	PPS	0-3%	Fine Loamy/Loamy Skeletal, Mixed Typic Rhodustalfs.	Horticulture Plantation
2.	Kharif Unirrigated	PPS	5-10% and 10-15%	Loamy Skeletal, Mixed Typic Ustorthents.	Silvi-Pasture
3.	Kharif Unirrigated	PPS	0-3%	Fine Loamy, Mixed Fluventic Typic Ustorpepts.	Conversion to Double Crop
4.	Kharif Unirrigated	PPS	5-10% and 10-15%	Loamy Skeletal/Coarse Mixed, Lithic/Typic Ustorthents.	Horticulture Plantation
5.	Kharif Unirrigated	PPS	5-10%	Fine Loamy/Loamy Skeletal, Mixed Typic Ustorpets/Rhodustalfs.	Intercropping System
6.	Kharif Unirrigated	Valley Fill	0-3%	Fine Loamy/Loamy Skeletal, Mixed Typic Ustorpets/Rhodustalfs.	Conversion to Double Crop
7.	Kharif Unirrigated	PPS	5-10% and 10-15%	Fine Loamy/Loamy Skeletal, Mixed Typic Rhodustalfs.	Horti-Pasture
8.	Kharif Unirrigated	PPM	3-5% and 5-10%	Coarse to Fine Loamy, Mixed Typic Ustorpepts.	Intercropping System
9.	Kharif Unirrigated	T/TC	5-10%	Loamy Skeletal/Mixed, Lithic/Typic Ustorthents. Loamy Skeletal/Mixed, Typic Ustroepts/Haplustalfs.	Agro-Forestry
10.	Kharif Unirrigated	RH	5-10% and 10-15%	Loamy Skeletal/Mixed, Lithic/Typic Ustorthents. Loamy Skeletal/Mixed, Typic Ustroepts/Haplustalfs.	Silvi-Pasture
11.	Kharif Unirrigated	PPS	0-3%	Loamy Skeletal/Mixed, Typic Ustroepts/Haplustalfs.	Agro-Horticulture
12.	Kharif Unirrigated	PPS	3-5%	Loamy Skeletal/Coarse Mixed, Typic Ustorthents.	Horticulture Plantation
13.	Kharif Unirrigated	PPS	3-5%	Loamy Skeletal/Mixed, Typic Ustroepts/Haplustalfs.	Horticulture Plantation
14.	Kharif Unirrigated	PPS	0-3%	Coarse to Fine Loamy, Mixed Typic Ustorpepts.	Conversion to Double Crop
15.	Kharif Unirrigated	PPS	3-5%	Loamy Skeletal/Fine Loamy, Mixed Typic. Rhodustalfs/Haplustalfs.	Horticulture Plantation
16.	Double Crop	PPS	0-3%, 3-5%	Fine Loamy/Loamy Skeletal, Mixed Typic Rhodustalfs. Fine Loamy/Loamy Skeletal, Mixed Typic Ustorpets/Rhodustalfs. Fine Loamy, Mixed Typic Ustorpets.	Agro-Horticulture

(*Contd.*)

(*Contd.*)

Sl. No.	*Existing Landuse/Cover*	*Hydro-Geomorphology*	*Slope*	*Soil*	*Recommended Landuse System*
17.	Double Crop	VALLEY FILL, PPM	0-3%, 3-5%	Fine Loamy/Loamy Skeletal, Mixed Typic Rhodustalfs. Fine Loamy/Loamy Skeletal, Mixed Typic Ustorpets/Rhodustalfs. Fine Loamy, Mixed Typic Ustorpets.	Existing Double Crop
18.	Fallow Land	PPS	0-3%	Coarse to Fine Loamy, Mixed Typic Ustorpepts.	Conversion to Double Crop
19.	Fallow Land	PPS	3-5%	Fine Loamy/Loamy Skeletal, Mixed Typic	Horticulture Plantation
20.	Plantation	PPS	3-5%, 5-10%	Loamy Skeletal, Mixed Typic Ustorthents.	Existing Plantation
21.	Degraded/Under-Utilised Forest	IRRESPECTIVE.	Irrespective.	Irrespective.	Gap Plantation
22.	Land With/Without Scrub	PPS	10-15%, 15-35%	Loamy Skeletal, Mixed Typic Ustorthents.	Silvi-Pasture
23.	Land With/Without Scrub	PPS	5-10%	Loamy Skeletal, Mixed Typic Ustorthents. Loamy Skeletal/Coarse Mixed, Lithic/Typic Ustortents. Loamy Skeletal/Mixed, Typic Ustroepts/Haplustalfs.	Horti-Pasture
24.	Land With/Without Scrub	PPS	10-15%, 15-35%	Loamy Skeletal, Mixed Typic Ustorthents. Loamy Skeletal/Coarse Mixed, Lithic/Typic Ustortents. Loamy Skeletal/Mixed, Typic Ustroepts/Haplustalfs.	Fodder and Fuel Plantation
25.	Land With/Without Scrub	PPS	3-5%	Loamy Skeletal/Coarse Mixed, Lithic/Typic Ustortents. Coarse to Fine Loamy Mixed Typic Ustorpepts.	Horticulture Plantation
26.	Land With/Without Scrub	PPS	10-15%, 15-35%	Loamy Skeletal, Mixed Lithic/Typic Ustortents. Fine Loamy/Loamy Skeletal, Mixed Typic Rhodustalfs.	Silvi-Pasture
27.	Barren Rocky/Stony Waste/Sheet Rock Area	PPS	10-15%, 15-35%	Loamy Skeletal/Mixed, Typic Ustroepts/Haplustalfs. Loamy Skeletal, Mixed, Lithic/Typic Ustorthents.	Fodder and Fuel Plantation
28.	Mining Area	RH	10-15%, 15-35%	Loamy Skeletal/Mixex, Typic Ustroepts/Haplustalfs. Loamy Skeletal, Mixed, Lithic/Typic Ustorthents.	Stabilisation of Mining Dumps
29.	Salt Affected Area	PPM	0-3%	Coarse To Fine Loamy, Mixed Typic Ustorpepts.	Salt Resistant Crops

The area where soil is salt effected, water potential is excellent, slope is gentle, salt resistant crops are proposed. The statistics of action plan unit is given in Table 10.6.

Table 10.6: Showing Alternate Landuse Classes and their Area

Sl. No.	*Alternate Landuse*	*Area (in Sq. Kilometre)*
1.	Gap plantation	9.85
2.	Horticulture	65.63
3.	Conversion of double crop	14.08
4.	Agro-forestry	4.39
5.	Horti-pasture	2.19
6.	Agro-horticulture	4.96
7.	Silvi pasture	1.94
8.	Intercropping system	6.93
9.	Existing water bodies	3.44
10.	Stablisation of mining dumps	0.88
11.	Existing double crop	7.34
12.	Salt resistant crop	1.06
13.	Existing plantation	0.09
14.	Built-up area	0.44
15.	Fodder cultivation	10.92
	TOTAL	10.92

The type study can be give required input for district administrator and other agencies for planning the best use of the available resources in order to improve the socio-economics conditions of the region. The action plan can assist in developing new policies and strategies for sustainable development.

REFERENCES

Burrough, P.A. (1986), *Principles of Geographical Information System for Land Resources Assessment*, Claredon Press, Oxford.

Chakraborty A.K, (1982), 'Assessment and Management of Land and Water Resources in Drought Prone Areas from Satellite Derived Data—an Indian Example', *Proceeding of the International Symposium of Remote Sensing of Environmentally First Thematic Conference: Remote Sensing of Arid and Semi-arid Lands*, Vol. 1, Cairo, Egypt, Publisher: Environmental Research Institute of Michigan, Ann, Arbor, Michigan.

Desai, Anjana, (1987), 'A GIS Based Study In Backward Regions of Gujarat', Land Resource Management, Abha Lakshmi Singh (Ed.), B.R. Publishing Cooperation, Delhi-92.

Dhinwa, P.S. (1993), 'Landuse Change Analysis of Bharatpur District Using GIS', *Photo Nirvachaka: Journal of the Indian Society of Remote Sensing*, Vol. 21, No. 2, September.

ESRI, (1992), *PC Overlay User Guide Version 3.4d*, ESRI, Inc. 380, New York Street, Redlands, Ca, USA.

HARSAC (1999), 'Integrated Study Through Space Applications for Sustainable Development in Gurgaon District,' *Haryana State Remote Sensing Application Centre (HARSAC)*, Department of Science and Technology, Haryana, CCS-HAU Campus, Hissar.

NRSA (1993), '*Technical Report: Evaluation of Landuse/Landcover Map Accuracy of IRS Multi-spectral Satellite Data,* ' National Remote Sensing Agency, Department of Space, Govt. of India, Balanagar, Hyderabad.

NRSA, (1995), '*Technical Guidelines: Integrated Mission for Sustainable Development*,' Department of Space, Govt. of India, Balanagar, Hyderabad.

Perumal, A., (1999), *Role of Space Systems in an Integrated Management Plan of Sustainable Development*, Integrated Survey Group, National Remote Sensing Agency, Department of Space, Govt. of India, Balanagar, Hyderabad.

Rao, Jayaraman and Chandrashekhar (1996), 'Organization of GIS Based Natural Resources Databases and Role of State Remote Sensing Centres,' *National Workshop on Application of Remote Sensing and GIS Techniques to Integrated Rural Development,* 14-15 June, Hyderabad.

Sahai, Baldev, (1993), 'Integrated Approach for Resource Development Using Remote Sensing—A Case Study Specific to Hill Areas', *Photo Nirvachak: Journal of the Indian Society of Remote Sensing*, Vol. 21, No. 2 September, 1993.

11

CAPABILITY CLASSIFICATION OF WASTELAND

MANU BAJPAI

Wasteland has been receiving a lot of attention in recent decades. The ever increasing population pressure and consequent demands for food, fuel and fodder have made it essential to look for additional land that could be utilised for satisfying some of those demands. The vast stretches and varieties of wasteland that exist throughout the length and breadth of the country are the areas, which attract attention as the potential lands that could be identified, developed and put to gainful uses. This study is an attempt to classify wasteland on the basis of their capability so that viable options of wasteland utilisation can be explored for the purpose of land use planning.

NATURE OF PROBLEM

Land comprising soil, water and associated plants and animals involving the total ecosystem is the most important among major resources of the country. Its efficient utilisation determines the pace of India's economic growth. Ill effects of development like destruction of forest cover, mismanaged irrigation, waterlogging, salinity, and unscientific agricultural practices are responsible for degradation of land resources. Adverse effects of environmental forces are also adding to the increasing menace of degradation.

Productivity of present agriculture land cannot be increased beyond certain limits, moreover, application of technology on a mass scale is limited in our country. In order to meet the demands of ever-increasing population and to achieve the cherished goal of sustainable development, reclamation and utilisation of wasteland are inevitable. Our country is facing severe problem of wastelands and if allowed to continue it may reach alarming heights. According to National Wasteland Development Board, eighteen per cent of Indian geographical area is categorised as wasteland. Evaluation of problem on district level shows that 148 districts of India are critically affected by the problem of wasteland. In Etawah district, which is the study area of the present work, more than 600 sq. km. of land are turned into different types of wastelands. These degraded lands are posing threats not only to environment and ecological security but also to sustainable socio-economic development. Various projects on saline soil and ravine land reclamation are running from last few years but the problem remains unchanged.

In the present work, wasteland capability classification is given special attention. The capability class determines the potential use of any piece of land and further helps in its utilization and planning. Such studies may help in reducing pressure of population on land and also to meet agricultural demand for the growing. In this respect it is desirable that we focus our attention to utilise wasteland and evaluate it on the basis of its potential so that intelligent decision can be taken to utilise it for different purposes and needs.

REVIEW OF LITERATURE

The available literature relevant for the present study is as follows: Patil assessed various geographical aspects of land use of Karla village of the Poona district and determined land potentials by using Benguni formula. Das and Das studied how environmental setting has created conditions for rational use of soils for agriculture with reference to land capability. Sharma and Sharma discussed the land capability classification in the Padrouna block of the Deoria district with a view to suggest optimum utilisation of land. They also suggested capability classification keeping in view for future conservation. The Soil Conservation Department, Govt. of India, 1970 has prepared "All India Standard Capability Classification Pattern" which evaluated soils for their susceptibility to erosion, soil-depth, drainage problems and other vital soil characteristics. Das and Bhattacharya suggested production variable as base for land capability classification and indicated that variation in land capability can be explained by the variation in yields. Rao and Vaidyanadhan presented the land capability classification of Krishna delta by studying landforms, soils and existing land use pattern with the aid of air photo interpretation. Rao evaluated the land suitability of the Madurai district with the help of remote sensing data. Factors which are considered by the author to evaluate land suitability classes are irrigation, cropping pattern and physical attributes, namely, wasteland, slope, landform, lithology, hydromorphic units, soils and land use. Madhuramma and Rao presented land capability classification of the Nellore district by using remote sensing data where land capability has been evaluated on the basis of physical characteristics. Young and Goldswith outlined the F.A.O. frame work of land evaluation by the application of soil management units. Pant, Joshi and Jalal adopted different techniques for determining land capability classification which is based on soil depth, slope, erosion intensity, water holding capacity, sunlight conditions and potential land use. Singh presented the land capability of Manipur valley by applying the techniques of rating the land by potential production units.

Studies on reclamation of wastelands are vast and diverse. Singh studied the economics and geography of agricultural land reclamation in the Koil tahsil of the Aligarh district. Jha presented the effective utilisation of different types of wasteland in the Birbhum district. The Academy of Science of USSR has discussed the application of drainage and other scientific reclamation measures for the saline soils. Sharma, Tripathi and Sharma described various reclamation measures for Usar land in the Sangrampur block of the Sultanpur district. Prasad and Yadav have prepared a list of questions and answers for the alkali and saline soil reclamation and have also pointed out water quality, drainage measure, cultivation, sowing time, nutrient requirements, control of pest and diseases for reclamation. Kumar and Panday determined the inherent characteristics of wasteland, causes and hazards for reclamation and controlling the destructive tendency operating over them. This work also provides necessary information required for conservation programme under which wasteland presently degraded by erosion can be improved.

Singh presented the possibilities of reclamation of cultivable wastelands which are being used as threshing grounds and sites for social and religious performance. Yadav and Padmaja attempted to reclaim Usar land of the Sultanpur district by correlating various morphometric, climatic, hydrogenic, pedogenic factors and presented reclamation structures on the basis of quantifying values. Yadav and Aggarwal provided measures to reclaim saline and alkali soils along with assessment of their potential. Shukla studied the eroded lands of the Yamuna and Chambal rivers of Etawah district and provided empirical structure for ravines reclamation.

The literature available on wasteland capability classification is scarce. However, few scholars have attempted in this direction. Yadav (1984) presented capability classification of wasteland in Sultanpur district considering various environmental and anthropogenic factors which quantify the extent of influence of these factors. In another study, Yadav (1987) discussed the conceptual framework of wasteland capability classification by defining reclamation cost as the suitable base for categorising capability classes. Kumar (1989) adopted integrated approach to wasteland capability determination for land use planning of Deoghar district. The author encoded environmental attributes, assigned capability value and gave them weighted and computed rating for each wasteland. The present work was an attempt to present an outline on the utilisation of wasteland

Even though capability classification is a precursor to land reclamation, we have found in the literature that more emphasis has been laid on reclamation studies. Most of the reclamation studies were done by government and related agencies, whereas individuals have not treated it as area of interest. Even though intensive literature is available on land capability classification, wasteland capability classification has not been covered extensively. The emphasis on wasteland reclamation for the purpose of utilisation can only be effectively planned when evaluation of wasteland potential is included in the studies.

Objectives of the Study

The study was conducted to fulfil following objectives:

(1) To show distribution of wasteland in Etawah district.
(2) To present wasteland capability classification by measuring the intensity of various physical and anthropogenic factors chosen for present study.

Study Area

The Etawah district lies in the south-western part of Uttar Pradesh in infertile soils of the Ganga-Yamuna doab. It extends between 26°21'N latitude to 27°01'N and 78°45'E to 70°45'E longitude (Fig. 11.1). It is bounded on the north by Mainpuri and Farrukhabad districts, on the east by Kanpur district, on the South by Jalaun district, on the west for some distance by Agra district and the rest by the State of Madhya Pradesh. The district lies entirely in the Ganga plain but its physical features differ considerably and are determined by the Yamuna and Chambal rivers.

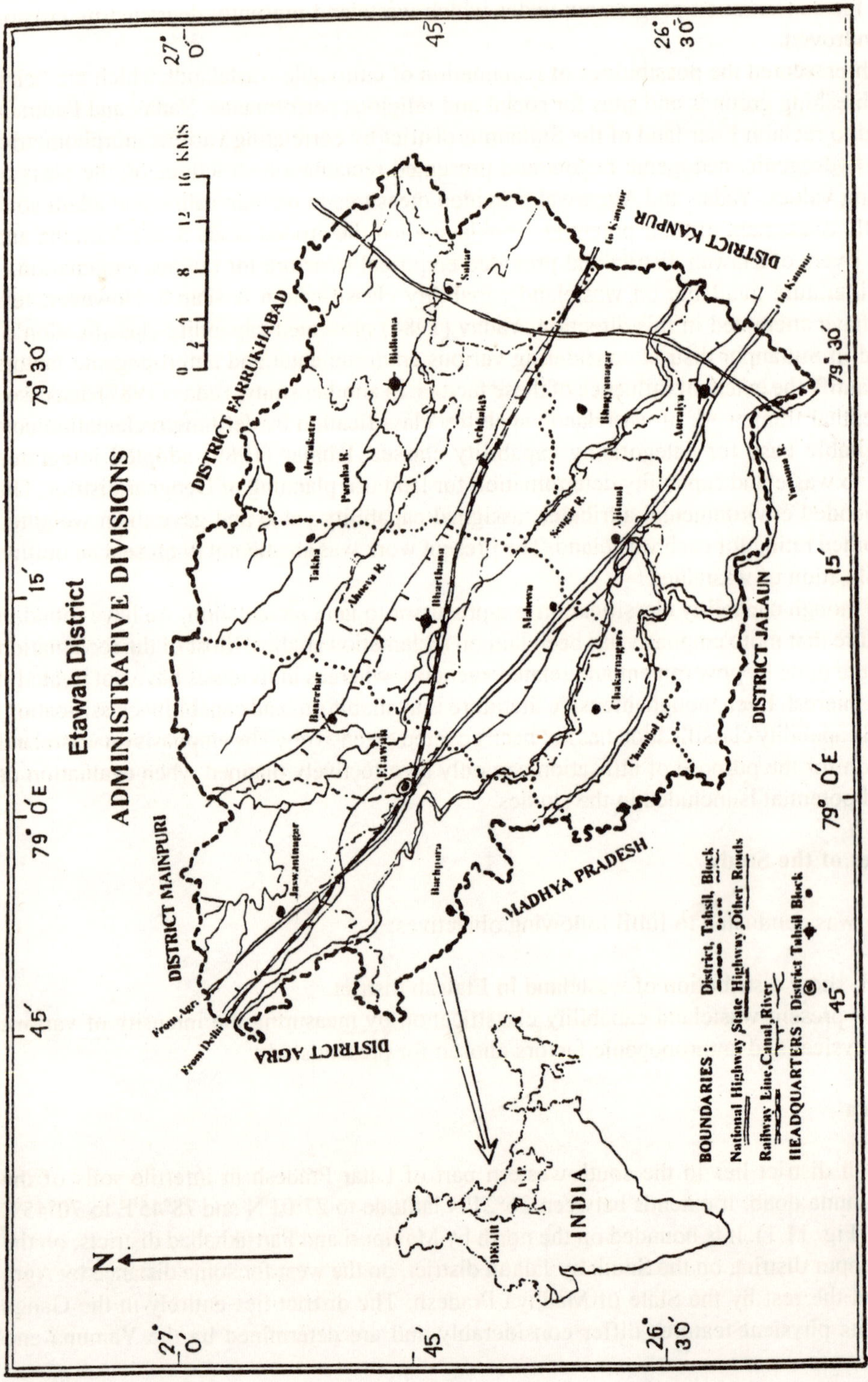
Etawah District
ADMINISTRATIVE DIVISIONS
DISTRICT MAINPURI
DISTRICT FARRUKHABAD
DISTRICT KANPUR
DISTRICT JALAUN
DISTRICT AGRA
MADHYA PRADESH
INDIA
N
0 4 8 12 16 KMS
BOUNDARIES :
District, Tahsil, Block
National Highway, State Highway, Other Roads
Railway Line, Canal, River
HEADQUARTERS District Tahsil Block
Etawah
Bharthana
Auraiya
Bidhuna
Jaswantnagar
Chakarnagar
Ajitmal
Takha
Basrehar
Mahewa
Sahar
Achhalda
Airwakatra
Barhpura
Bhagyanagar
Chambal R.
Yamuna R.
Sengar R.
Ahneya R.
Puraha R.
to Kanpur
From Agra
From Delhi
79°0'E
79°30'
15'
45'
27°0'
26°30'

Fig. 11.1

The present work chooses district as a unit of study because the present work is structured in such a way for which a district is more suitable as a study area. Secondly, as a district is a recognised unit of socio-economic development in various government plans, it is more practical to correlate wasteland with this aspect. All the Government policies and projects on wasteland are applicable on administrative units which make easier to collect data and information.

The Etawah district aptly represents the problem of wastelands faced by U.P. Few districts of Uttar Pradesh are so severely affected with the problem of wastelands as the Etawah district. Six-hundred sq. km. of area is badly affected with various types of wastelands. Ravine land affected the southern region of the district. Saline soil is evenly distributed in northern part of the district. Waterlogged land, upland and degraded forestland are other types of wasteland found in the district. Only the middle portion of the district which is sandwiched between northern saline soil and southern ravine soil is agriculturally productive region. This spatial distribution of wasteland makes Etawah peculiar and complex as compared to other districts of the state. Northern adjoining districts of Etawah like Mainpuri. Etawah, and Fatehpur are mainly saline affected, whereas southern adjoining districts like Bhind are ravine affected districts. The other main objective to choose the district is that the district is poor in socio-economic development, which can be correlated with poor utilisation of land and distribution of wasteland. The area has a potential for agricultural development and suitable measures for utilisation of wasteland will serve as input into the economic betterment of the area as a whole. The Etawah district is now divided into two districts. This study is conducted by taking into account the older administrative unit of Etawah district.

DATA SOURCES

Data for present study were collected from both primary and secondary sources. General information related with people, agriculture and socio-economic aspects were collected from District Census Handbook. Village map of the district was obtained from Primary Census Abstract. Detailed geological map on 1:250,000 scale of the district was collected from Geological Survey of India, Lucknow. Soil maps of the district were collected from National Bureau of Soil Survey and Land Use Planning, New Delhi which also provide information on soil characteristics of the district. Slope map was derived from soil map of the district, which had detailed slope information. Land erosion map of the district was obtained from National Bureau of Soil Survey and Landuse Planning.

Information on natural regions was obtained from Census Atlas of Uttar Pradesh. Wasteland maps of the district on 1:50,000 scale was obtained from Remote Sensing Application Centre, Lucknow. The areas under different wastelands were calculated by maps. Ravinous soil map on 1:250,000 scale of the district was collected from Geological Survey of India, Lucknow. Saline soil map was collected from National Remote Sensing Agency, Hyderabad. Data on various determinants used for capability classification were collected from secondary sources. Rainfall data was collected from Indian Meteorological Department. Data on groundwater level and Ph value of groundwater were collected from Central Groundwater Commission, Lucknow. Sodicity, electrical conductivity of soil and surface soil drainage information were collected from National Bureau of Soil Survey and Land use Planning, New Delhi and Remote Sensing Application Centre, Lucknow. Some valuable information was also obtained from Usar Land Reclamation Project, sponsored by World Bank. Canals and tubewell data were obtained from Deputy Director, Agriculture Office, District Etawah.

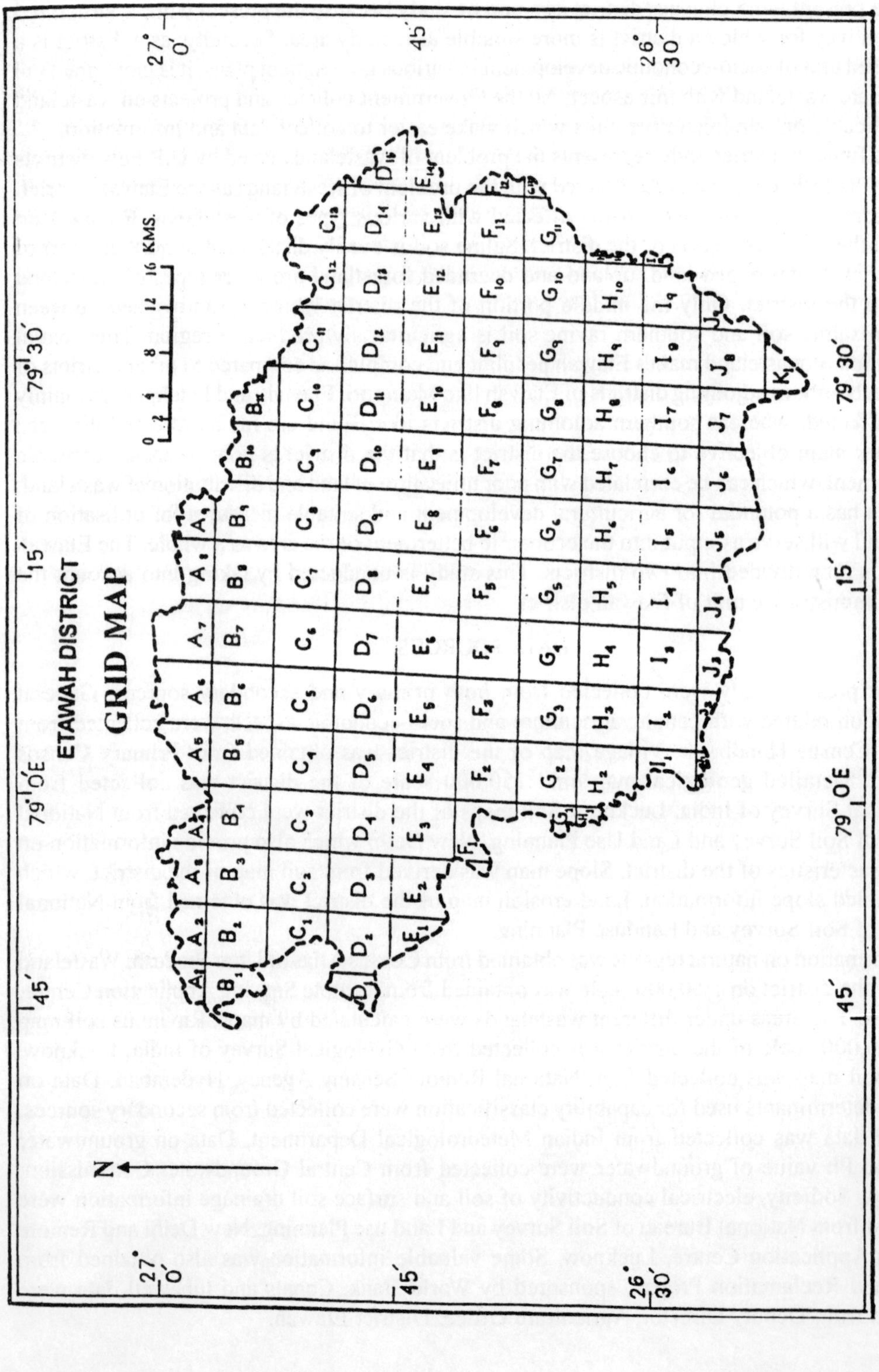
ETAWAH DISTRICT
GRID MAP
N
0 2 4 8 12 16 KMS.
79° 0E
79° 30′
45′
15′
27° 0′
26° 30′

Fig. 11.2

METHODOLOGY

Distribution, identification and assessment of wasteland was done by using maps of 1:50.000 scale which were obtained from Remote Sensing Application Centre. Demarcation of wasteland is the basic task of this research as accurate analysis of other objectives depends upon this work. Wasteland was mapped for every administrative unit, that is, from village to district level. Demarcation of wasteland is done on the basis of criteria given by National Wasteland Development Board. Maps of different types of wastelands on district, tahsil and block level have been prepared. Attempt has also been made in present study to correlate wasteland with various physical and socio-economic conditions of area. Wasteland area of each and every village is calculated with the help of graph method and curve meter since reliable data is lacking.

Capability classification of wasteland was done on the basis of various parameters:

(a) Sodicity of the soil, (b) surface soil drainage, (c) electrical conductivity of the soil, (d) land erosion, (e) slope, (f) water table depth (in dry season), (g) electrical conductivity of groundwater, (h) mean annual rainfall, (i) number of tube wells (state and private tube wells), and (j) length of canals.

The study was conducted on two levels, namely, district and village. Capability classification on district level is the difficult task because of a selection of unit for a micro-level analysis. Few conceptual and practical problems arise in the selection of micro unit for capability classification such as:

(a) Villages cannot be selected for capability classification as a unit because there are 1563 villages and it is impractical for present purpose.
(b) All types of wastelands found in Etawah are so evenly distributed in patches of varied size that it is difficult to categorise these wastelands into grid. Size of patches ranges from less than one hectare to more than 1000 hectares.

Therefore, to overcome these problems whole of the Etawah district was divided into grids. Distribution of wasteland in various grids was properly demarcated. Intensity of various factors taken for the capability classification was quantified and assessed grid by grid. On the basis of that wasteland in different grids was grouped into four categories which are:

(a) Possibilities of utilisation are better
(b) Possibilities of utilisation are moderate
(c) Possibilities of utilisation are poor
(d) Possibilities of utilisation are very poor

Where the intensity of all the ingredients are unfavourable, it implies lack of appropriate conditions for utilisation of wasteland. In the areas where intensity of factors is favourable it means wasteland can be utilised through various inputs and efforts. This is the basic approach behind the categorisation of wasteland into various capability classes. Capability classification is done on the basis of severity of wasteland. Some factors which do not show significant variations in the study area e.g. depth and texture of soil etc. are excluded from capability classification.

DISTRIBUTION OF WASTELAND

The Etawah district is prone to the problem of wasteland. More then 1/10th of its area is subjected to the wasteland (Fig. 11.3). Though, the distribution of wasteland is not uniform in the district, yet all the 14 blocks of the district constitute one or the other type of wasteland. The Barhpura block has more than 22 per cent of its total area under ravenous land. Besides this, blocks like Bidhuna, Chakarnagar and Auriya have 15 per cent to 20 per cent of their area under wasteland. Ajitmal is the least affected block of the district having only 4 per cent of its area under wasteland (Fig. 11.3).

Rest of the blocks constitute about 10 per cent of its area under wasteland. Out of total 54604.39 hectares of wasteland in Etawah district, more that 15 per cent, i.e. 8255 hectares are located in Auriya block. Bidhuna and Chakarnagar have more than 23 per cent of the total wasteland of Etawah district which is 5459 and 6044 hectares, respectively. The Ajitmal block is relatively less prone to wasteland.

The wasteland problem of the district is depicted more clearly if number of villages affected by wasteland in each block is analysed. Out of total 1543 villages of Etawah district around 925 (60 per cent of total villages) are affected by wasteland. In Chakarnagar block no village is observed to be free from wasteland. Similarly Bidhuna, Takha, Sahar and Barhpura blocks, most of the villages are prone to wasteland. The Ajitmal and Mahewa blocks are found relatively free from wasteland. (Table 11.2). Most of the villages in the district are affected by Saline wasteland. The villages affected by salinity are comparatively more in number than the villages affected by ravenous land though the area under ravanious land is comparatively more.

TYPES OF WASTELAND

In Etawah district five types of wastelands are found. These are: ravines land, saline land, waterlogged land, degraded forest land and upland with or without scrub. Detailed distribution of these lands is given below.

Ravinous Wasteland

Ravinous land, cover 2,68,551.8 hectare area of the district which is 51.75 per cent of the total wastelands and 6.13 per cent area of the district. Yamuna, Chambal and Kunwari river basin covers 23,551.8 hectares of land. Major proportion of this ravinous land occurs in Auriya tahsil. At most of the places, these ravine patches are located near the reserved or protected forest and are infested with grasses, babul or other thorny bushes, and scrubs as cover.

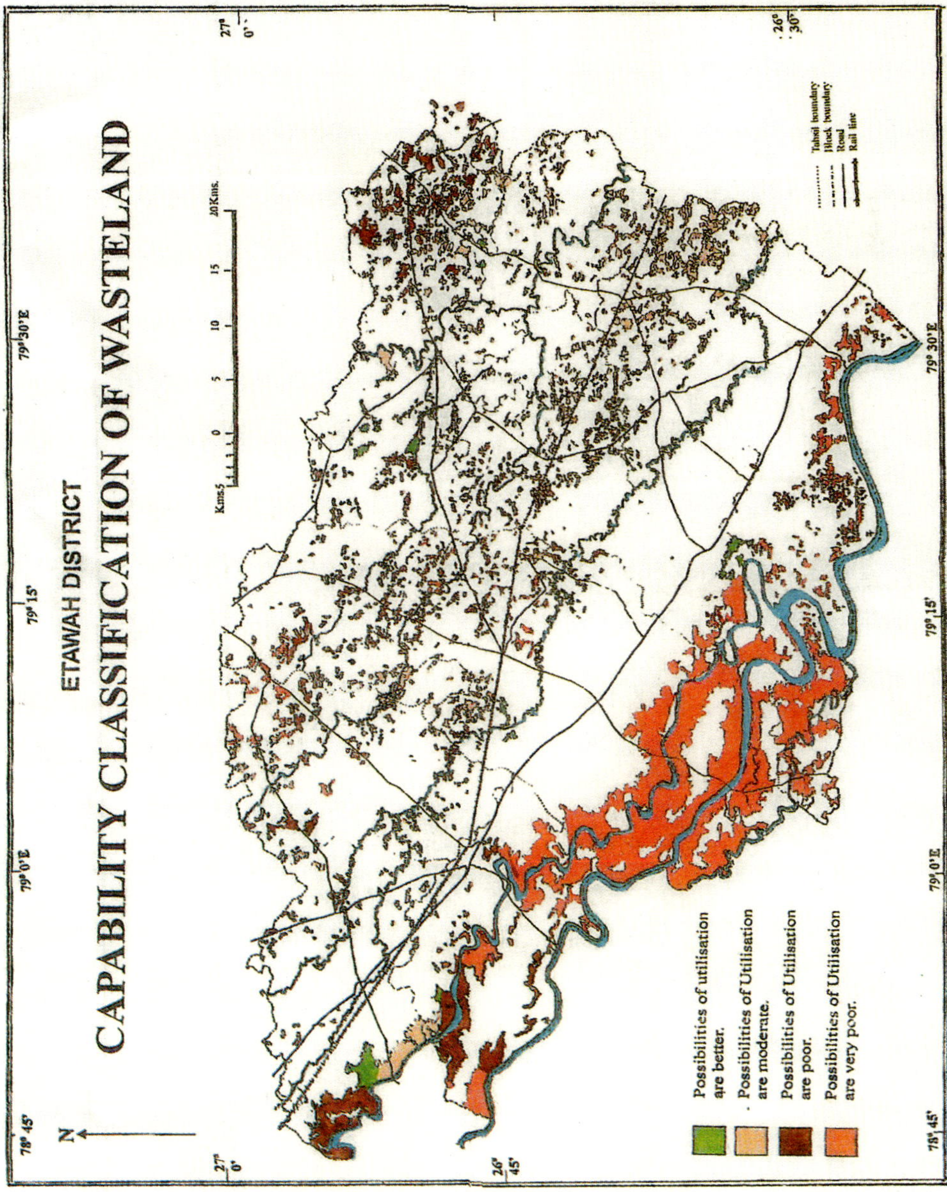
ETAWAH DISTRICT
CAPABILITY CLASSIFICATION OF WASTELAND
N
Kms.5 0 5 10 15 20Kms.
78° 45'
79° 0'E
79° 15'
79° 30'E
27° 0'
26° 45'
26° 30'
Possibilities of utilisation are better.
Possibilities of Utilisation are moderate.
Possibilities of Utilisation are poor.
Possibilities of Utilisation are very poor.
Tahsil boundary
Block boundary
Road
Rail line

Fig. 11.3

Table 11.1: Distribution of Wasteland in Etawah District, 1990

(Area in Hectares)

Sl. No.	*Block*	*Area (in hectares)*	*Type of Wasteland*					*Total*
			Ravines	*Saline*	*Water-logged Land*	*Degraded Forestland*	*Upland*	
1.	Jaswantnagar	38800	2472	1786	10	–	–	4250
2.	Bidhuna	31000	39.5	4648.2	810	0.15	–	5459.35
3.	Basrehar	37500	39.5	3046.1	378.5	–	–	3464
4.	Bharthana	25900	–	2534.2	29	–	12	2575.2
5.	Takha	27500	–	2218	492	–	–	2710
6.	Mahewa	32400	2514	246	–	32	–	2792
7.	Chakarnagar	37200	5911	–	–	133	–	6044
8.	Airwakatra	22400	–	1760.3	240	–	–	2000.3
9	Sahar	28400	–	2263	108	164	635	2598.3
10.	Auriya	41400	7877.6	229	–	199	–	8225.6
11.	Ajitmal	20700	753.5	30	–	30	–	813.5
12.	Bhagyanagar	27600	–	3406.6	60	129	15	3676.5
13.	Achalda	28200	–	2731.5	5	25	96.5	2858
14.	Bharhpura	32900	6918.2	128.54	–	67	–	7107.54
15.	Etawah Dist. Total	432600	26551.8	2132.5	2132.5	778.3	758.3	54460.4

Source: Primary data calculated with the help of maps of wasteland acquired from Remote Sensing Application Centre, Lucknow, 1990.

The ravine wasteland on the basis of the magnitude of the problem of soil erosion can be demarcated into two categories: (a) Areas where the problem is more acute and which are dominated by a network of deep gullies and wild ravines. (b) Areas where the gullies and ravines are not very deep and dense and which occupy a comparatively narrow zone along the river banks. The eroded land of the catchment areas of the river Yamuna, Chambal and Kunwari come in the first category where as those of the river Sengar, Ahaniya, Puraha, and Arind in second category. Block where the largest area is affected with ravinous wasteland is Auriya followed by Barhpura but in terms of percentage of the total block area, Chakarnagar is covered with 16 per cent of the total area by ravine wasteland. Bidhuna, Bharthana, Takha, Airwakatra, Sahar Achalda blocks are found free from ravine land. Eight blocks of Etawah with total 216 villages are affected with ravinous wasteland. Auriya block has the largest number of villages affected with wasteland followed by Barhpura, Chakarnagar and Mahewa.

LAND AFFECTED BY SALINITY/ALKALINITY

Saline wasteland is the second largest category in the district which cover 25,024.34 hectares accounting for 40.42 per cent of total wastelands and 5.78 per cent of the total district area. The northern part of the district, especially in north-east, is severally affected with salinity. All the blocks excepting Chakarnagar are affected with salinity with varying degree of magnitude. Proportions of saline soil vary from 14.99 per cent total area of Bidhuna to 75 per cent of the total area of Mahewa block.

Table 11.2: Villages affected by Wasteland in Etawah District: 1990

Block	*Total Villages (in No.)*	*Number of Villages Affected with Wasteland*					
		Saline	*Ravines*	*Water-logged*	*Degraded Forestland*	*Upland*	*Total*
Jaswantnagar	131	50	15	1	–	–	66
Bidhuna	105	84	–	24	1	–	109
Basrehar	142	74	4	13	–	–	83
Bharthana	82	57	–	3	–	1	57
Takha	70	54	–	14	–	–	55
Mahewa	114	8	22	–	3	–	33
Chakarnagar	38	–	31	–	7	–	38
Sahar	96	72	–	5	8	6	75
Auriya	192	12	73	–	4	–	92
Ajitmal	112	5	15	–	3	–	19
Bhagyanagar	134	80	5	6	4	4	83
Achalda	118	69	–	1	1	14	74
Barhpura	94	10	51	–	4	–	63
Airwakatra	115	63	–	15	–	–	78
Etawah Dist. Total	1543	638	216	82	39	25	925

Source: Primary data generated with the maps of wasteland obtained from Remote Sensing Application Centre. Scale of maps is 1:50,000, 1990.

The salt affected patches occur in all the four tahsils. These wastelands occur mostly along and around the Arind river where large patches can be seen between Sahar distributary, Kansa distributary and also south-west of Bela town in Bidhuna tahsil. Other salt affected wasteland patches occur along and around the Etawah branch of the Ganga canal starting from village Kanchousi upto the end of Bidhuna. The western portion of Bidhuna tahsil is also infested with the salt affected land between Bidhunas-Umrain and Bidhuna Kudarkot Road. Large patches also occur on the north of Kudarkot town. A cluster of big and small salt affected lands occur along and around Kuthipur and Phaphund distributary. The land affected by salinity/alkalinity in Bharthan tahsil is mostly located in the central part of the tahsil. It is also found towards the western portion of the tahsil extending from north to south. Some patches are found between Purahanadi, Bansok distributary, Aheniya Nadi, Boharpur distributary, Balunda distributary and Sengar river. In the eastern portion of the tahsil, small salt affected patches are seen between Sengar river and Puraha Nadi. In Etawah tahsil, small patches of salt affected land occur along the Sengar river on the right bank. Large salt affected patches occur in the northern portion the tahsil along Bansok distributary, Etawah branch of the Ganga canal and along the Aheniya Nadi. In Auriya Tahsil, salt affected patches are localised mostly towards the eastern part of the tahsil. These patches occur between Sengar river, Mangalpur distributary and Phaphund town.

Block level pattern shows that saline soil varies from 14.9 per cent in Bidhuna to 0.75 per cent in Mahewa block. Chakarnagar is the only block of the district which is free from saline soil. Blocks which are severally affected with saline soil are Bidhuna, Bhagyanagar and

Basrehar. Area of saline soil in these blocks are 4649.2, 3406.6, 3046.1 hectares respectively. Least affected blocks are Ajitmal, Mahewa Auriya and Barhpura. In Etawah district, 638 villages are affected with saline soil of which highest number is found in Bidhuna (84) followed by Basrehar (74), Bhagyanagar (80), Sahar blocks. Chakarnagar is the only block where no village is affected with saline soil. Least number of villages affected with saline soil is found in Ajitmal, Mahewa and Barhpura.

WATERLOGGED AND MARSHY LAND

This is the third category of wasteland in the district and which occurs on 2132.5 hectare of land accounting for 4.72 per cent of total wasteland and 0.48 per cent of the total district area. The northern portion of the district particularly north-east is affected with this type of wasteland. Nine blocks out of 14 are affected with waterlogged land. Waterlogged land in these blocks varies from 5 hectares in Achalda to 810 hectares in Bidhuna. Mahewa, Chakarnagar, Auriya, Ajitmal, Barhpura are the blocks which are observed to be totally devoid of this type of wasteland. Generally, there seems to be a close relationship between drainage density and waterlogged land. In case of Etawah district it is found that blocks having more drainage density have no waterlogged land. Interestingly, Chakarnagar, Auriya, Barahpura where Chambal, Yamuna river flow from west to east are devoid of waterlogged land. The obvious reason for not having waterlogged land along these major rivers is the depth of river bed and slope of surface towards river. In the north of the district, no major river flows and the slope is level to level plain. As such, waterlogged land is found along canals.

Tahsil level analysis shows that waterlogged land mostly occurs in Bidhuna tahsil. Some waterlogged lands in this tahsil are found near and around the Lower Ganga Canal (West Allahabad Branch). Some patches are also found in the northern part of Bharthana and Etawah tahsils. Thus, marshlands occur on both the sides of Etawah-Fatehgarh Road and have water hyacinth growing on it. Bidhuna block has the highest number of villages affected with waterlogged land is found in Takha. Airwakatra also have some share of villages affected with waterlogged land. Blocks, namely, Ajitmal, Auriya, Chakarnagar, Mahewa have no villages affected with waterlogged land. Total eighty two villages of Etawah district are found affected with waterlogged and marsh land.

DEGRADED FORESTLAND

This pertains to the land notified under the Forest Act, in which denudation of vegetative cover has resulted in degraded land. The vegetative cover has less than 20 per cent. This category occupies 779.15 hectares of land which is 1.67 per cent of the total wasteland and 0.180 per cent of the total land area of the district. In terms of forest area it covers 3.37 per cent of the reserved/protected forest in the district. In Etawah tahsil the distribution of this category is restricted along left bank of the river Yamuna near Bathi, Nagla Salahi, Nagla Tor, Sarongpur and Bulokipur village and Purha, Dangl and Bohri village on the right bank. In tahsil Bharthana, these lands occur on the south of Yamuna near Miholi and Khitriti villages and between Chambal and Kunwari.

In Bidhuna, tahsil degraded forests occur near village Jamlapur and Singh Purwa Mana on the north of Arind river whereas in tahsil Auriya, these lands occur near Debiapur, and also along Yamuna river near Akbarpur, Asta and Nagla Banadas village. Nine blocks of Etawah district are affected with degraded forestland, which varies from 199 hectares of Auriya to 0.15 hectares of Bidhuna. Degraded forestland is the highest in Auriya followed by Sahar, Bhagyanagar and Chakarnagar blocks. Jaswantnagar Basrehar, Bharthana, Takha and Airwakatra are totally devoid of degraded forest land. Total, thirty-nine village of the district are affected with this type of wasteland. Auriya, Sahar, Chakarnagar block have the highest number of villages affected by degraded forestland.

UPLAND WITH OR WITHOUT SCRUB

This type of wasteland occurs on 758.3 hectare of land, which are 174 per cent of the district area and 0.44 per cent of the total wasteland area of the district. These wastelands occur as small patches and are scattered along the river Sengar and its tributaries. These are generally eroded peripheral lands. In these uplands, scrubs and grasses in scattered patches are found. This wasteland is found in 4 blocks of Etawah district. The Sahar block has the largest proportion covering an area of 635 hectares whereas Bhartana has the least coverage of 13 hectares of land. Bhagyanagar and Achada are other blocks reported to be affected with 25 villages, the highest numbers of affected villages are found in Achalada and Sahar. Blocks, namely, Auriya, Ajitmal, Barahpura, Airwakatra, Jaswantnagar, Bidhuna, Bidhuna, Takha, Mahewa and Chakarnagar are free from uplands.

CAPABILITY CLASSES OF WASTELAND

The wasteland of Etawah district is classified into four capability classes on the basis of factors mentioned earlier. Wasteland capability class I includes those areas where aggregate value of choosen factors are above 90, this class is represented by green colour on capability map (Fig. 11.4). Only 7.2 per cent (i.e. 3967.5) of the total wasteland area is covered by this category, found mostly in few separate areas lying along Yamuna and in north central part of the district. Capability class II is represented by light orange colour on capability map which represents the areas with aggregate value between 80-90. This capability class includes saline and waterlogged lands which are found scattered in the district lying north of Sengar river. Seventy per cent of wasteland in north-eastern part of the district belongs to wasteland capability class II.

Table 11.3: Area Under Wasteland Capability Classes

Sl. No.	*Categories*	*Aggregate Point*	*Area (in hectares)*	*Area (in %)*
1.	I	Above 90	3967.5	7.2
2.	II	90-80	14851	27.7
3.	III	80-70	13195	24.1
4.	IV	Below 70	23325	42.7
	TOTAL		54601	100

Source: Area calculated from the wasteland capability map prepared by author.

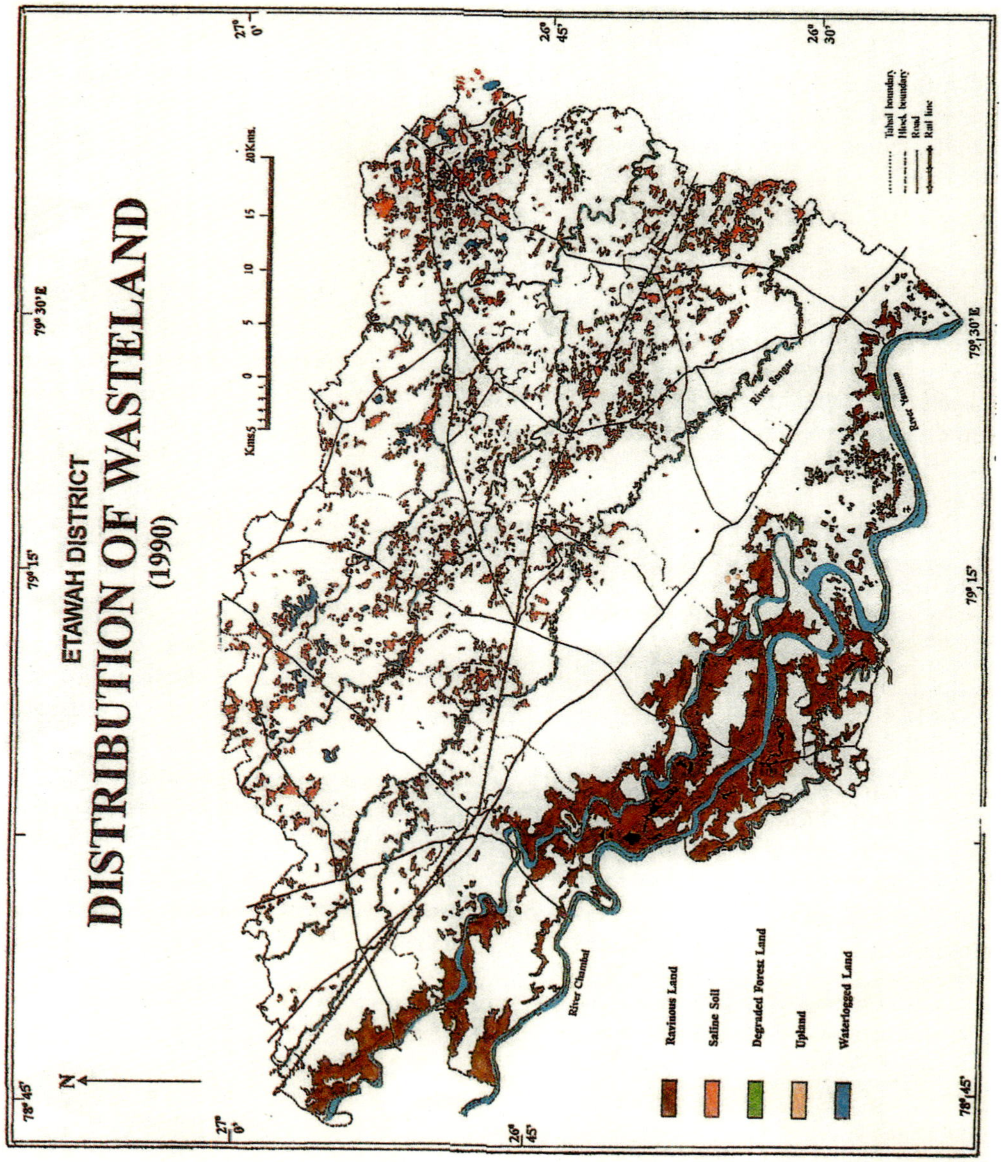
ETAWAH DISTRICT
DISTRIBUTION OF WASTELAND
(1990)
N
Kms.5 0 5 10 15 20 Kms.
Ravinous Land
Saline Soil
Degraded Forest Land
Upland
Waterlogged Land
Tahsil boundary
Block boundary
Road
Rail line
River Chambal
River Sengar
River Yamuna
78°45'
79°15'
79°30'E
27° 0'
26° 45'
26° 30'

Fig. 11.4

Wasteland capability class III, shown by brown colour. It covers about 24 per cent of the total wasteland in the district. Areas of this class are found scattered in north of the river Senger. Large stretch are also in north-eastern part of the district. Most of the saline and waterlogged lands represent this capability class.

Wasteland capability class IV, represented by red colour, constitutes more than 40 per cent of the total wasteland of the district. Possibilities of utilisation are very poor in the wasteland of this class. Ninety per cent of ravine land are included in this class. Which are located in southern part of the district along river Yamuna and Chambal. Some saline areas of central and northcentral part of the district also represent this class of wasteland. Table 11.4 shows block level distribution of wasteland capability classes.

Wasteland Capability Class I

Wasteland of this class covers 3967.5 hectares of the total wasteland of the district. The possibilities of utilisation are better in the areas under this class. These wastelands are found in all blocks except Takha, Chakarnagar, Bhagyanagar and Achalda blocks (Fig. 11.5). Along the Yamuna river, two patches of 208 hectares are found in Ajitmal block. These are degraded forest land, where initial forest has been degraded.

Along the river Yamuna, in Jaswantnagar block, area of 935 hectares is included in this class. This area is mostly ravinous land, but the intensity of ravines is not very deep, thus, possibilities of reclamation are better. We also find numerous patches of class I wasteland along Sengar river in Mahewa block covering an area of 120 hectares. These patches are covered mainly by saline soil. The region surrounding Kharaghpur town in Airwakatra block also contains class I category of wasteland which roughly covers 500 hectares of land.

Wasteland under class I is mostly free from land erosion, except few patches along river Yamuna. Slopeout, electrical conductivity of groundwater are also favourable in this class while sodicity and electrical conductivity of soil is relatively high. Other factors mentioned earlier make the wasteland under class I relatively favourable for utilisation. These lands can be utilised for agricultural purposes with little inputs and efforts. Wasteland belonging to this class mostly represents saline soil. Using saline resistant crops along with suitable inputs can show favourable results in agriculture.

Etawah and Bidhuna tahsil have greater proportion of wasteland under capability class I. All the blocks, except Takha, Chakarnagar, Bhagyangar and Achalda have capability class I land. Jaswantnagar block has 935 hectares of class I wasteland while Airwakatra block have 600 hectares of land where possibilities of utilisation are better.

Wasteland Capability Class II

Capability class II includes wasteland where possibilities of utilisation are moderate. It consists 14851 hectares of land. Wasteland under this class is found in areas drained by Sengar, Purha and their tributaries. The National Highway 2 separates the wasteland of capability classes II and IV. District north of NH_2 contains wasteland of capability class-II while in south of NH_2 capability class IV is predominant. Eighty per cent of the wasteland of class II are found in central-eastern part of the district, which is mostly saline and waterlogged land.

Table 11.4: Block Level Distribution of Wasteland Capability Classes

Block	*Wasteland Area (in hectares)*	*Wasteland Capability Classes*							
		Class I Area		*Class II Area*		*Class III Area*		*Class IV Area*	
		(In hect.)	*(In %)*	*(In hect.)*	*(In %)*	*(In hect.)*	*(In %)*	*(In hect.)*	*(In %)*
Jaswantnagar	4250	935	22	1147.5	27	2167.5	51	–	–
Bidhuna	5459.35	163.78	3	1910.77	35	3384.79	62	–	–
Basrehar	3464	242.48	7.1	1558.8	45	866	25	796.785	23
Bharthana	2575.2	128.77	5	1158.98	45	849.92	33	486.96	17
Takha	2710	–	–	596.2	22	1084	40	1029.8	38
Mahewa	2792	223.36	8	83.76	3	67.52	6	2317.36	83
Chakarnagar	6044	–	–	–	–	–	–	6044	100
Airwakatra	2000.3	600.04	30	500.07	25	900.13	45	–	–
Sahar	2598.3	519.66	20	2078.64	80	–	–	–	–
Auriya	8255.6	165.11	2	–	–	–	–	8090.40	98.1
Ajitmal	813.5	207.44	25.5	16.27	2.5	20.33	2.5	568.75	70
Bhagyanagar	3676.5	–	–	3235.3	88	367.65	10	72.53	2
Achalda	2858	–	–	1714.8	60	1114.62	39	28.58	1
Barhpur	7107.54	781.82	11	852.90	12	2274.41	32	3909.14	55
TOTAL	54601	3965.5		14851		13195		23325	

Source: Area calculated from the wasteland capability map prepared by the author.

In these areas groundwater depth is favourable, land erosion in minimum and surface soil drainage is good. Agriculture is possible in these wastelands and salinity resistant crops can be use for cultivation. Dense network of canals and tubewells adds to the possibility of utilisation, because canal water is good for leaching process in the saline soil wasteland of western and south-western part of Bharthana town represents class II land. Some ravine soils in eastern part of the district are also included in this category. Although salinity resistant crops can be grown in capability class II wasteland, but social forestry of economically profitable tree species would be more beneficial and practical because salinity is relatively higher than capability class I and utilisation of wasteland would require larger input.

In Bidhuna tahsil more than 45 per cent of total wasteland belongs to the category of moderate possibilities of utilisation. In Auriya and Etawah tahsil, 25 per cent and 23.9 per cent of total wasteland fall under the capability class II. All the blocks except Auriya and Chakarnagar have wasteland under class II. In Bhagyanagar, Sahar, Bidhuna, Achalda block more than 90 per cent of the total wasteland represent wasteland of capability class II. More than 40 per cent of the total wasteland representing a capability class II is in Jaswantnagar, Barahpura, Takha, Bharthana blocks.

Wasteland Capability Class III

Possibilities of utilisation are poor in capability class III. This class covers an area of 13195 hectares of land, i.e. 24.1 per cent of the total wasteland of the district. Wasteland of this class is found in three regions lying in eastern, central and north-western parts of the district. The State highway 21 separates these three regions. Mostly, saline and waterlogged land constitute wasteland of this class, but in north-western part near Yamuna and Chambal river ravine lands are included in capability class III. Wasteland belonging to this category located east of State highway 21 is unique because very high groundwater level which is responsible for high sodicity and salinity of the soil. Salitity more than 30 ec/dsm^{-1} is found in this particular region. Similarly, sodicity above 40 esp makes wasteland unsuitable for any type of agricultural activity. Groundwater level in this region is very high which increase the water salt content in the surface soil during dry season. This region looks like hard light grayish pan in dry season. Only few sodicity resistant tree species can be grown here. Moreover, lack of agricultural infrastructure and low economic prosperity of this region hinder the reclamation measures. User Land Reclamation Project sponsored by World Bank have surveyed 3 villages in this region but results are not favourable. It was observed that very high salinity and sodicity makes any reclamation strategy futile, as land reverts back to its original state within 2 to 3 months. Saline soil expands very rapidly, therefore, it is necessary to check its growth to adjoining areas.

Wasteland of capability class III, lying in central part of the district, represents more or less similar conditions, as mentioned for the region east of SH. 21, but severity of impact is relatively less. Land erosion, slope, surface drainage and groundwater depth are unfavourable in this region. Besides this, the region also lacks the irrigation sources because of the infavourablity of these factors.

Block level picture shows that all the blocks have presence of wasteland of capability class III, except Chakarnagar, Sahar and Auriya. In Bidhuna, Basrehar and Barahpura more than 70 per cent of wasteland falls in capability class III.

Wasteland Capability Class IV

Possibilities of utilisation are very poor in the wasteland capability class IV. This class covers an area of 23325 hectare (i.e. 42.7%) of the total wasteland of the district. Thus, this class covers majority of wasteland in the district. Major portions of this class are found along the river Yamuna and Chambal in southern part of the district. More than 90 per cent of wasteland of capability class IV consists of ravines soil. Rest of the wasteland includes saline soil of central and north central part of the district. Ravines soil has a rugged topography, which makes its utilisation almost impossible. Severe land erosion and slope are main drawbacks of wasteland of this class. Wasteland of this class is isolated region of the district. The depth of ravines is more than 15 meters in certain areas which manifest the severity of land erosion. Groundwater is below 20 meters. It is also below 30 meters in a few areas. Land erosion is more than 70 tonnes per hectare per year. In the wasteland of capability class IV slope is more than 15 per cent in most area.

Development of infrastructure for agriculture and economy is very poor, which further diminishes the possibility of utilisation. Though salinity, soil texture is relatively better but severity of other phenomena, such as land erosion, slope makes their favourability insignificant and renders their utility. Deforestation in these areas also leads to severe land erosion. It is imperative to provide vegetation for this region so that land erosion can be controlled.

Wasteland of capability class IV is not found in Bidhuna, Airwakatra and Sahar. In Chakarnagar block wasteland of capability class IV covers 6044 hectares while in Auriya block it covers 8090.78 hectares. These figures show that block in southern part of the district contains major share of wasteland under capability class IV.

Overall, picture shows that most of the wastelands located in central part of the district have better possibility of utilisation because intensity of factors in this part are the least. Infrastructure development, economic conditions also increase the possibilities of utilisation of wasteland. Severity of wasteland increases as we move from central to northern and southern part. In the south, severity of wasteland increase because of intense land erosion, moderate low slope, groundwater depth. The importance of this capability classification lies in the fact that reclamation and utilisation can be applied according to the possibilities of utilisation of each class. Thus, the classification gives a fair idea for the planner as to how to implement the reclamation and utilisation plans.

The analysis portrays the acuteness of wasteland problem in the Etawah district. Wasteland covers 12 per cent of the district area mostly by saline and ravines land. Others type of wasteland cover small proportion in comparison of saline and ravenous land. Wastelands located in central part of the district have better possibility of utilisation because intensity of factors in this part are the least. Infrastructure development, economic conditions also increase the possibilities of utilisation of wasteland. Severity of wasteland increases as we move from central to northern and southern part. In the south, severity of wasteland increase because of intense land erosion, moderate low slope, groundwater depth. The importance of this capability classification lies in the fact that reclamation and utilisation can be applied according to the possibilities of utilisation of each class. Thus, the classification gives a fair idea for the planner as to how to implement the reclamation and utilisation plans. It should very clearly be understood that like other development programmes the wasteland development programme is the programme of the people, by the people and for the people, and they expect a greater contribution to the development process and objectives. Need of hour is not to take isolated efforts but to create a system which should involve every one including common people to government for action role to play. Then only we can succeed to mitigate this problem and judiciously utilise our land resources

REFERENCES

Academy of the Science of U.S.S.R., (1958), The Application of Drainage in the Reclamation of Saline Soils, Translated from Russian, Mascow.

Aggarwal, R.K., (1980), "Physio-chemical states of soils under khejri prospopis cineraria linn" in Mahn, H.S. and Sexsena, S.K. (ed.), *Khejri in the Indian Desert*, Monograph No. 11, Jodhpur, CAZRI: 31-36.

Aggarwal, R.R., (1957), "Delhi soils can be reclaimed", *Indian Farming*, 7 (9): 24-48.

Bennet, H.H., (1939), Soil Conservation, McGraw Hill Company, New York.

Bhumbla, D.R. and Khare, A., (1984), Estimate for Wasteland in India, Society for Promotion of Wasteland Development, New Delhi.

Clark, A.N. Longman Dictionary of Geography, Longman, London.

Davidson, D.A., (1992), The Evaluation of Land Resources, Longman, London.

Uttar Pradesh District Gazetters, Etawah. (1981), Government of Uttar Pradesh, Lucknow: XI.

Gautam, N.C., (1983), "Satellite Remote Sensing Techniques for Natural Resource Survey" in Singh L.R. and *et.* (ed.). *Environmental Management, Allahabad Geog. Society*: 177-181.

Haussen Builler, R.L., (1963), Principles of Soil Science, Orient Longman, Bombay.

Govinda Rajni, S.V., (1966), Soil Survey Manual, All India Soil and Landuse Survey Organisation, New Delhi.

Jha, B.N., (1965), "Land Capability Classification of the Kosi Region", *Geog. Outlook*, (4): 76-77.

Jha, V.C., (1984), "Wasteland Types and their Effective Utilisation in Birbhum District", *The Deccan Geographer*, (41): 231-242.

Kumar, A. Panday. R.N. and Singh, S.P., (1983), "Wasteland a Case for Environmental Management" in Singh L.R. (ed), *Environmental Management*, Allahabad Geographical Society, Allahabad: 229-242.

Kumar, A. and Pandey, R.N. (1989), Wasteland Management in India, Ashish Publishing House, New Delhi.

Kumar, A., (1989), "Integrated Approach to Wasteland Capability Determination in Landuse Planning: A Case Study of Deoghar District of Bihar", in Singh (ed), *Problem of Wasteland and Forest Ecology in India*, Ashish Publishing House, New Delhi: 89-115.

Mohammad, N. (1988) "Wastelands in India: Some Issues and Problems" in Sharma S.C. (ed.) *Utilisation of Wasteland for Sustainable Development in India*, Concept Publishing Company, New Delhi: 57-65.

Mukherjee, A.B., (1971), "Geographical Patterns of Changes in Agricultural Waste Lands in Uttar Pradesh", *Geographical Review of India*, 2(33): 120-132.

National Remote Sensing Agency. (1991), Wasteland Atlas of India, Hyderabad.

N.R.S.A., (1986), Manual of Procedure for Wastelands Mapping Using Remote Sensing Techniques, Department of Space, Govt. of India.

Prasad, R. and Yadav, J.S.P., (1969), Alkali Soil Reclamation, Question and Answers, Central Soil Research Institute, Karnal, India.

Remote Sensing Application Centre, (1990), District Wasteland Report: Etawah district, Lucknow.

Russell, E.J., (1950), Soil Conditions and Plant Growth, Longman Green and Co. London.

Shafi, M., (1968), "Problems of Wasteland in India", *The Geographer*, (15): 1-15.

Shafi, M., (1969), "Can India Support Five Times her Population?", *Science Today*, (2): 21-27.

Sharma, H.S., (1968), "Land-capability classification of the lower Chambal Valley" Proc. Symp. Land use in Developing Countries, Aligarh Muslim University:173-179.

Sharma, H.S., (1980), Ravines Erosion in India, Concept Publishing Corporation, New Delhi.

Sharma, S.C. Tripathi, D.N. and Sharma, M.L., (1988), "Usar land Reclamation Measures in India with special Reference to Sangrampur Block, District Sultanpur, U.P" in S.C. Sharma (ed), *Utilisation of Wasteland for Sustainable Development in India*, Concept Publishing Company, New Delhi: 285-302.

Shuckla, P.N., (1988), "Yamuna and Chamal Rivers Eroded Land and its Reclamation in Etawah district, U.P" in S.C. Sharma (ed): *Utilisation of Wasteland for Sustainable Development in India*, Concept Publishing Company, New Delhi: 359-368.

Singh, A.L., (1978), Economics and Geography of Agriculture Land Reclamation, B.R. Publishing Corporation, New Delhi.

Singh, B., (1962), "Distribution of Cultivable Waste and Possibilities of its Reclamation in Chakia, Banaras", *The National Geographer of India*, 1(7): 59-72.

Singh, R.L., (1971), Indian Regional Geography, National Geographical Society of India, Varanasi.

Soil Survey Manual and U.S.A. Development of Agriculture (1954), A Manual on Conservation of Soil and Water, Dept. of Agriculture, U.S.A.

Spate, O.H.K., (1954), India and Pakistan, Mathuen, London.

Stamp L.D., (1954), The Land of Britain, it Use and Misuse, Longman, London.

Storie, T.E., (1933) "An Index for Rating the Agricultural Value of Soils" California Experimental Station Bulletin, No. 556: 221-229.

U.S.A. Dept. of Agriculture., (1951), Soil Survey Manual, U.S.A.

Yadav, H.R., (1987) "Agroforestry Management Systems in wastelands" in Workshop Proceeding of Agroforestry for Rural Needs: II, Khosala P.K and Khurana, D.K, I.S.T.S., Solan, India.: 108-22.

Yadav, H.R., (1987), "Suitability Classification of the Wasteland", in H.R Yadav (ed.) Wastelands: Diognosis and Treatment, Concept Publishing Company, New Delhi: 93-98.

Yadav, H.R., (1984), Genesis and Utilisation of Wasteland, Concept Publishing Company, New Delhi.

Yadav, J.S.P., and Aggrawal, R.R., (1959) "Dynamics of soil changes in the reclamation of saline, alkali soils of the Indo. Gangatic alluvium", *Journal of Indian Society of Soil Sciences* (7): 77-86.

Young, A. and Goldemith, P.F. (1977), "Soil Survey and Land Evaluation in Developing Countries: A Case Study of Malawai", *Geographical Journal*, (43): 407-438.

12

REMOTE SENSING FOR SOIL RESOURCE INVENTORY OF PART OF GUNDAR BASIN, VIRUDNAGAR DISTRICT, TAMIL NADU

A. VELMURUGAN AND T.P. SWARNAM

INTRODUCTION

The knowledge on soil resources of an area is vital for any developmental activities. The characteristics of soils in a region are heterogeneous and classification of soils based on their characteristics facilitates optimum land use planning and better soil resource management. Recently the satellite remote sensing has emerged as an extremely valuable tool for identification on natural resources for different aspects of sustainable development. An attempt has been made to characterize the soils of part of Gundar basin in Tamil Nadu through satellite remote sensing and classification as per USDA soil taxonomy.

MATERIALS AND METHODS

The study area covers parts of Arupukottai and Tiruchuli taluks of Virudunagar district, Tamil Nadu (Fig. 12.1). The area consists of laterite and alluvium of Pleistocene period, basic dykes of Precambrian and Gondwana periods with marine stratification of lower cretaceous period. The physiography is nearly undulating to level lands with gentle slope. The mean annual temperature ranges from 24.3°C (min) to 34.5°C (max.). The annual rainfall is 900 mm.

Landsat 5 TM FCC Geocoded imageries on 1: 50,000 (58 K2 and 58 K6) along with SOI toposheets were visually interpreted based on interpretation keys like tone, texture, size, pattern, mottles and association. Different imagery interpretation units were identified and sample strips were selected so as to cover all the interpretation units. Soil pedons were examined in each unit and morphological characteristics were described (Anonymous 1971). Soil samples were collected and analysed for physico-chemical properties by adopting standard procedures (Jackson 1973). Finally soils were classified taxonomically as per Soil Taxonomy (Soil Survey Staff 1998). Soil map of the area was also prepared.

RESULTS AND DISCUSSION

Remote sensing is an excellent tool for identification of physiographic units and delineating relief features and landforms which in turn assists in the identification of soil distribution pattern. The characterisation and mapping of soils using remote sensing technique is based on the spectral characteristics of soils. Reflectance of soils is very much influenced by the morphological and chemical composition, soil structure, surface roughness, particle size and organic matter content. The spectral response data depicted in the form of various image interpretation elements along with the site characteristics from a reliable database for identification of soils and their classification.

The landsat 5 TM, FCC Geocoded imageries on 1: 50, 000 scales were visually interpreted based on photo-elements like tone, texture, size, shape, pattern, aspect, association etc., along with SOI toposheets and geological map of the study area. Based on the interpretation keys, 50 interpretation units were identified. The physiography, land use and soil type in each interpretation unit were identified during field visit.

FIELD WORK

Sample strips were selected so as to cover all interpretation units. During field work the major soil types in each interpretation unit were fist identified visually. Limited ground checks were also conducted in the study area to verify the results of the similar interpretation unit in outside the sample strip area. The units possessing similar soil types were brought under one group and thirty-two units thus finally arrived from the mapping units for preparing soil maps. Thirteen representative pedons were examined in detail and characterised for morphological, physical and physico-chemical properties.

MORPHOLOGICAL PROPERTIES

The morphological features of pedons were given in Table 12.1. There was an observable difference in soil colour, which could be related to either physical or chemical or biological properties of the soils. The soil colour varied from different shades of gray through yellowish red to red. Lime, iron concretions and mottles were observed in most of the pedons. In two of the pedons examined soickensides were also observed. In most of the pedons, the surface horizons were brown in colour. In other pedons it was either gray or red to yellowish red under dry condition (Pedon Sevalpatti, Tiruchuli, Udayanendal, Palayampatti, Erumaikulam and Ramanujapuram). The various shades of gray observed in pedons Sevalpatti, Tiruchuli and Thimmapuram might be due to the presence of organic matter or absence of free iron oxides and reduced condition. The red to yellowish red colour was observed in pedons Udayanendal, Palayampatti and Ramanujapuram. The red colour is due to presence of free iron oxides and well-oxidized conditions (Menon and Mariakulandai 1957).

Lime concretions were observed is Swevalpatti, Tiruchuli, Taraganendal, Mandriodai, Erumaikulam and Ramanujapuram pedons. The presence of lime concretions might be due to the nature of parent materials (Kaushal *et. al.* 1986). In subsurface horizons of Pillayarkulam pedon distinct black (10 YR 2/1) coloured mottles and yellowish red (5 YR 5/8) mottles were observed in Avarangulam pedons. The occurrence of mottles might be due to fluctuating water table creating alternate oxidized and reduced conditions within the pedon. Similar observations were also made by Diwakar and Singh (1994) in alluvial soils of Bihar.

Table 12.1. Morphological Properties of the Soils

Horizon	*Depth (cm)*	*Soil Colour Dry*	*Moist*	*Texture*	*Structure*	*Consistency*	*Boundary*	*Concretions*	*Mottles*
Sevalpatti									
A_{1pk}	0–27	10 YR 2/2	10 YR 2/2	Scl	3 msbk	dh,mfi,ws,wps	cs	very small round to irregular lime conca	–
B_{21k}	27–60	–	10 YR 2/2	Scl	m–3csbk	mvfi,ws,wps	cs	very small round to irregular lime conca	–
B_{22k}	60–98	–	10 YR 3/1	Scl	3 csbk	mfi,ws,wp	cw	very small round to irregular lime conca	–
B_{23k}	98–124	–	10 YR 3/1	Scl	3m–csbk	mfi,ws,wp	gw	few, large, round to irregular lime conca	–
B_{31k}	124–144	–	10 YR 3/2	Scl	3 csbk	mfi,ws,wp	dw	few, large, round to irregular lime conca	–
B_{32k}	144–150	–	10 YR 3/3	Scl	2 csbk	mfi,ws,wps	–	few, large, round to irregular lime conca	–
Tiruchuli									
A_p	0–18	10 YR 5/3	10 YR 4/3	Sl	m–2sbk	dvh, mf	gw	–	–
B_{21k}	18–42	10 YR 3/1.5	10 YR 3/2	Scl	2 msbk	mfi, ws, wp	cs	very few powdery lime conca.	–
B_{22k}	42–84	–	10 YR 3/2	Scl	3 csbk	mvfi, ws,wp	cw	very few, small & large powdery lime conca.	–
B_{23k}	84–119	–	10 YR 3/2	Scl	3 csbk	mfi,ws,wps	gw	small & large irregular lime conca	–
B_{31k}	119–141	–	10 YR 4/1	Scl	2 csbk	mfi,ws,wps	gw	common, round to irregular lime conca	–
B_{32k}	141–161	–	10 YR 4/2	Scl	m – 1 fgr	mfr,wss	–	dominant, round to irregular lime conca	–
Thimmapuram									
A_p	0–9	5 YR 5/6	5 YR 4/6	Sl	m–1fgr	dh,mfr	cs	–	–
B_{21w}	9 to 26	5 YR 5/6	2.5 YR 4/8	Scl	2 csbk	dh,msfi,wss	gw	few round, irregular quartz & very fine round iron concretions	
B_3	26–62	5 YR 5/6	2.5 YR 4/8	Cl	m–1 gfr	dsh,mfr	–	dominant, small & large irregular quartz & few round iron conca.	
Udayanendal									
A_p	0–17	10 YR 5/3	10 YR 4/3	Sl	m–3csbk	dh,msfi,wss	cw	–	–
B_{21}	17–34	10 YR 4/3	10 YR 4/3.5	Scl	m–3cbk	dvh,msfi,wss,wp	cw	–	–
B_{22}	34–62	–	10 YR 3/3	Scl	m–3csbk	dsh,msfi,wss,wp	as	–	–
B_{23}	62–98	–	10 YR 3/3	Scl	m–2csbk	dsh,msfi,wss,wp	aw	–	–
B_{24}	98–120	–	10 YR 3/3	Scl	m–2csbk	dsh,msfi,wss,wp	gw	–	–
C_1	120–142	10 YR 6/3	10 YR 5/4	S	M – 1 fgr	dl,mfr	–	–	–

(*Contd.*)

(Contd.)

Horizon	Depth (cm)	Soil Colour Dry	Moist	Texture	Structure	Consistency	Boundary	Concretions	Mottles
Palayampatti									
A_p	0–15	5 YR 4/6	5 YR 3/6	Ls	m–2csbk	dh.mfr,wss	gw	–	–
B_w	15–39	2.5 YR 3/6	2.5 YR 3/6	Scl	m – 1fgr	dl,mfr	dw	few, round to irregular small and large iron conca	–
Taraganendal									
A_p	0–20	10 YR 4/2	10 YR 4/4	Scl	m – fgr	dsh,mfr,wss	gw	very few round to irregular small and large iron conca	–
C_{1ca}	25–110	loosely cemented lime concretionary nodules							
Pillayarkulam									
A_{1p}	0–20	10 YR 4/3	10 YR 3/3	Scl	m–2csbk	dh,msfis,wss	cw	–	–
B_{11}	2054	10 YR 5/6	10 YR 6/4	Scl	m–3csbk	dvh,mfi,ws,w ps	cw	few, round, small iron conca	fine, few distinct black (10 YR 2/1') mottles
B_{12}	54–102	2.5 YR 6/4	2.5 Y 6/4	Scl	3cbsk	dvh,mfi,ws,w ps	dw	few, round, small iron conca	Fine, few distinct black (10 YR 2/1) mottles
B_{13}	102–134	–	10 YR 4/3	Scl	m	mvfi,ws,wps	Dw	Very few, small, round to irregular iron conca.	Comon, fine black mottles (10 YR 2/1) few, fine indistinct very dark brown (10 YR 2/20 mottles)
B_{14}	134–150	–	10 YR 4/3	Scl	m	msfi, wss	–	–	few, fine indistient very dark brown (10 YR 2/2) mottles
Mandriodai									
A_p	0–13	10 YR 5/3	10 YR 4/3	C	2 csbk	dh,msfi,ws,	cs	very few irregular to round, small & large lime conca	–
B_{2k}	13–36	10 YR 4/3	10 YR 4/2	C	3 csbk	wps	cw	" " " "	–

(*Contd.*)

Horizon	Depth (cm)	Soil Colour Dry	Moist	Texture	Structure	Consistency	Boundary	Concretions	Mottles
C_{1k}	36–70	10 YR 5/3	10 YR 4/3		m-1 csbk	dvh,mfi,ws,wp		very few irregular to round, small & large lime conca	
C_{2k}	70–87	–	–		powdery	vs			
C_{3k}	87–150	–			kankar			dominant, small, round to irregular lime conca	
Kalattur									
A_p	0–20	10 YR 4/3	10 YR 3/3	Sl	m-1mgr	dvh,mfr,wss	cs		
A_{12}	20–46	10 YR 5/8	10 YR 5/6	Scl	m-2msbk	dh,mfr,wss	gw		
C_1	46–79	–Laterite –							
C_2	79–108	–Laterite –							
C_3	108+	–Laterite_							
Avarangulum									
A_{1p}	0–12	10 YR 4/3	10 YR 4/3	Ls	m-1 fgr	dh,mfr	as	–	–
B_{21}	12–39	7.5 YR 5/6	7.5 YR 4/6	Scl	m-3 csbk	dvh,mfi,ws,wp	cs	–	–
B_{22t}	39–57	7.5 YR 5/6	7.5 YR 5/6	Scl	m-2csbk	s	cs	indistinct clay and iron cutans	–
B_{23t}	57–90	7.5 YR 5.5/4	7.5 YR 5/6	Scl	2csbk	dvh,msfi,wss	cw		–
B_{31}	90–123	7.5 YR 5/6	7.5 YR 5/6	Scl	m-1csbk	dsh,msfi,wss	gw	–	thick clay and iron cutans
B_{32}	123–150	7.5 YR 5/4	7.5 YR 5/8	Scl	m-2csbk	dsh,msfi,wss dsh,msfr,wss		few, small and large irregular iron conca few, small and large irregular iron conca	common, coarse distinct Mottles
Erunakulam									
A_p	0–20	10 YR 5/3	10 YR 4/3	Ls	m-1fge	dsh,mfr	cs	very few lime conca	–
A_{12}	20–42	10 YR 4/3	10 YR 4/4	Sl	m-2csbk	dvh,mfre	cw	–	–
A_{13}	42–54	10 YR 5/4	10 YR 4/4	Sl	m-3msbk	dh,msfi,wss	gw	very few, small round to	–
C_1	54–70	–	–	–	–	–	aw	irregular iron concretions indurated iron	–
C_2	70–98	–	–	–	–	–	–	concretionary nodules in the form of lateric lateric	–

(*Contd.*)

(*Contd.*)

Horizon	Depth (cm)	Soil Colour Dry	Moist	Texture	Structure	Consistency	Boundary	Concretions	Mottles
Perumalpatti									
A_p	0–10	2.5 YR 5/8	5 YR 4/6	Ls	m–1fgr	dh,mfr	cs	–	–
B_{21t}	10–32	2.5 YR 3/6	2.5 YR 3/4	Scl	2–3 csbk	dvh,msfi,wps	as	–	–
B_{22t}	32–54	2.5 YR 3.5/6	2.5 YR 3/6	Scl	2csbk	dvh,msfi,ws,	gw	–	–
B_{23t}	54–70	2.5 YR 4/6	2.5 YR 4/6	Scl	2csbk	wps	dw	very few, quartz and iron concretions	–
B_3	70–90	2.5 YR 4/6	2.5 YR 4/6	Cl	m–1fgr	dvh,msfi,ws,wps dsh,mfr	gw	very few, quartz and iron concretions	–
C_l	90–146	–	–	–	–	–		conglomerate containing quartzite, feldspar, kaolinite clay	
Ramanujapuram									
A_{1p}	0–15	10 YR 3/1	10 YR 3/1	C	1msbk–1fge	dh,msfi,ws,wps	cs	few round to irregular lime conca	–
A_{12k}	15–42	10 YR 3/1	10 YR 3/1	C	3csbk	dh,mvfi,ws,wp	cs	few round to irregular lime conca	–
A_{13kss}	42–65	–	10 YR 3/2	C	3cabk	mvfi,ws,wp	gw	few round to irregular lime conca	–
A_{14kss}	65–110	–	10 YR 3/1	C	3ccbk	mvfi,ws,wp	gw	few round to irregular lime conca	–
C_{1k}	110–135	–	–	Scl	–	–	–	dominant small and large irregular lime conca	–

The clay and iron oxide cutans were observed in subsurface horizons of Perumalpatti pedon. The transportation of clay by water through the larger voids in the soil and their deposition on the walls at a time when water left the voids result in the formation of clay cutans. It was also supported by laboratory analysis, which revealed increased clay content in subsurface horizons. The presence of Slickensides was noted in Sevalpatti and Ramanujapuram pedons indicating the high swell-shrink potential of the soil (Challa and Gaikwad, 1986).

PARTICLE SIZE DISTRIBUTION

The results of particle size distribution (Table 12.2) revealed wide variation in the clay content of different soils. It varied from 8.1 to 54.6 per cent. The clay content increased with depth in Palayampatti, Kallatur and Ramanujapuram pedons, while in other pedons it increased up to intermediate horizons and decreased thereafter. The coarse and fraction varied from 4.1 to 60.8 per cent in surface layers of all the soil pedons.

PHYSIO-CHEMICAL PROPERTIES

The physico-chemical properties of the soils were given in Table 12.2. The soil reaction varied from acidic to alkaline. The soils of Sevalpatti, Tiruchuli, Pillayarkulam, Mandriodai and Ramanujapuram were alkaline in reaction with pH values ranging from 7.9 to 10.4 indicating the high saturation or presence of free $CaCO_3$ or the dominance of calcium and magnesium ions in the exchange complex (Buol *et. al.* 1980). The highest pH value of 10.4 was recorded in Mandriodai indicating high degree of alkalinity and presence of appreciable amounts of exchangeable sodium. The acidic soil reaction was observed in soils of Thimmapuram, Udayanendal, Kallatur and Perumalpati (pH 4.9 to 6.4). Where as in other pedons neutral soil reaction was noted with pH values ranging from 6.8 to 7.2

The surface horizons of Sevalpatti, Tiruchuli, Taraganendal, Mandriodai and Ramanujapuram pedons registered comparatively higher CEC [24.12 to 40.2 cmol (+)/kg soil] than others. The nature of clay minerals shoed to have a profound influence on the CEC (Cation Exchange Capacity) of different soils. The surface horizons of Udayanendal and Pillayarkulam pedons showed CEC values varied from 20.05 to 21.54 cmol (+)/kg soil, in others it varied between 9.21 to 14.12 cmol (+)/kg soils. A significant and positive correlation recorded between CEC and clay content (r = 0.82022**) supported the results (Diwakar and Singh 1994 and Sahu *et al.* 1990).

The present base saturation indicates the degree of leaching and is used as a diagnostic criterion for classifying the soil order. The Sevalpatti, Tirupatti, Tiruchuli, Udayanendal, Taraganendal, Pillayarkulam, Mandriodai and Ramanujapuram pedons recorded relatively higher base saturation values (8.16 to 96.6 per cent), while in other pedons it varied from 49.4 to 78.9 per cent. The low base saturation in these soils could be attributed to low pH and CEO values. The significant, positive correlation obtained between BSP pH (r = 0.66794*) and BSP and CED (r = 0.690*) confirmed the above inference. In most soils the exchangeable complex was dominated by Ca^{2+} followed and Mg^{2+} followed by Na^+ and K^+ ions.

Table 12.2: Physical and Physico Chemical Properties of the Soils

Hirizon	*Depth (cm)*	*Soil Texture (Fine Earth Fraction) Clay %*	*Silt %*	*Sand %*	*pH*	*EC (dSm^{-1})*	*$CaCO_3$ (%)*	*CEC Cmol (+)/kg Soil*	*Exchangeable Cations Cmol(+)/kg soil* Ca^{2+}	Mg^{2+}	Na^{2+}	K^{2+}	*NSP*	*O.C %*
Sevalpatti														
A_{1pk}	0-27	32.0	8.4	59.6	8.38	0.56	7.0	34.92	15.70	12.28	4.25	0.42	92.2	0.28
B_{21k}	27-60	36.2	10.6	53.2	3	0.42	6.8	35.79	14.85	12.06	4.72	0.51	90.0	0.21
B_{22k}	60-98	37.3	13.6	49.1	8.3	0.56	5.0	39.07	16.23	15.09	3.98	0.51	91.6	0.17
B_{23k}	98-124	38.6	13.2	48.2	8.3	0.61	5.1	41.08	18.50	16.01	4.57	0.58	96.5	0.12
B_{31k}	124-144	35.4	13.8	50.8	8.2	0.95	5.0	32.05	13.50	11.98	3.93	0.64	93.5	0.12
B_{32k}	144-150	34.4	18.1	47.5	8.1	0.62	6.0	30.09	11.98	8.72	4.29	1.12	86.7	0.06
Tiruchuli														
Ap	0-18	16.1	18.6	65.3	8.3	1.38	5.0	24.17	12.57	5.04	2.98	0.19	86.0	0.25
B_{21k}	18-42	22.9	20.1	57.0	8.2	1.55	4.3	31.85	15.78	8.79	3.15	0.33	88.1	0.24
B_{22k}	42-84	26.9	20.6	52.5	8.2	1.82	4.8	33.85	17.88	9.44	3.41	0.33	90.3	0.23
B_{23k}	84-119	23.6	21.4	55.0	8.2	1.83	7.5	34.94	17.78	9.94	4.44	0.27	92.7	0.22
B_{31k}	119-141	18.3	21.9	59.8	8.3	1.82	6.5	31.58	15.11	7.96	4.68	0.24	88.6	0.20
B_{32k}	141-161	17.0	23.0	60.0	8.0	2.33	6.0	25.94	15.01	4.26	4.39	0.20	91.9	0.15
Thimmapuram														
A_p	0-9	14.6	15.4	70.0	6.0	–	9.21	3.55	1.61	0.16	0.16	0.46	62.7	0.20
B_{21w}	9-26	24.6	24.6	50.3	5.0	–	8.07	2.61	1.71	0.23	0.23	0.23	59.2	0.32
B_3	26-62	16.0	16.0	57.0	5.2	–	8.82	2.60	1.07	0.35	0.35	0.34	49.4	0.20
Udayanendal														
A_p	0-17	19.6	20.5	59.9	6.8	0.07	1.8	21.54	9.65	3.06	0.27	0.16	61.0	0.18
B_{21}	17-34	21.5	20.1	58.4	6.4	0.13	1.5	20.88	12.39	3.62	0.34	0.13	78.9	0.16
B_{22}	34-62	23.5	20.6	55.9	6.3	0.14	1.4	25.63	13.93	4.77	0.45	0.13	81.5	0.06
B_{23}	62-98	20.1	24.4	55.5	7.2	0.15	2.5	23.52	15.49	5.02	0.56	0.13	90.1	0.05
B_{24}	98-120	16.9	18.6	64.5	7.3	0.16	4.6	20.97	14.05	3.63	0.73	0.30	89.2	0.03
C_1	120-142	1.9	17.6	80.5	7.5	0.06	4.6	13.20	8.32	2.56	0.50	0.13	87.2	0.02
Palayampatti														
A_p	0-15	14.3	14.6	71.1	7.0	0.26	–	13.23	9.59	8.02	0.27\	0.16	83.4	0.18
B_w	15-39	26.0	23.0	51.0	6.8	0.17	–	23.10	10.44	5.37	0.47	0.23	71.4	0.31
Taraganenedal														
Ap	0-25	27.1	2.–0	70.9	7.2	0.10	35.0	32.90	17.52	10.67	2.00	0.40	92.3	0.22
Pillayarkulam														
A_{1p}	0-20	28.0	6.0	66.0	8.0	0.30	3.0	20.05	15.08	3.04	0.22	0.26	92.8	0.23
B_{11}	20-54	27.6	10.9	61.5	8.1	0.22	4.5	15.05	11.12	2.53	0.63	0.26	96.6	0.11
B_{12}	54-102	28.6	13.1	58.3	8.2	0.38	4.3	16.79	11.16	35.98	1.89	0.22	93.7	0.02
B_{13}	102-134	23.6	17.3	59.1	8.2	0.36	4.6	13.71	6.03	4.53	1.67	0.29	89.0	0.93
B_{14}	134-150	18.8	18.7	62.5	8.3	0.36	4.0	10.05	4.50	2.08	2.08	0.23	88.0	0.02
Mandriodai														
A_p	0-13	54.6	24.3	21.1	10.3	1.74	35.1	33.56	12.07	9.28	15.05	0.30	95.2	0.15
B_{2k}	13-36	40.6	19.9	39.5	10.4	2.62	39.0	37.73	13.67	11.34	10.97	0.46	96.6	0.13

(Contd.)

(*Contd.*)

Hirizon	*Depth (cm)*	*Soil Texture (Fine Earth Fraction) Clay %*	*Silt %*	*Sand %*	*pH*	*EC (dSm^{-1})*	*$CaCO_3$ (%)*	*CEC Cmol (+)/kg Soil*	*Exchangeable Cations Cmol(+)/kg soil* Ca_{2+}	Mg_{2+}	Na_{2+}	K_{2+}	*NSP*	*O.C %*
Kalattur														
A_p	0-20	13.6	22.8	63.6	5.3	0.05	–	10.05	3.57	2.54	0.27	0.29	66.3	0.31
A_{12}	20-46	19.6	22.4	58.0	4.9	0.05	–	10.05	3.05	2.03	0.26	0.26	52.6	0.22
Avarangulam														
A_{1p}	0-12	8.1	17.3	74.6	5.9	0.52	–	11.40	3.61	3.55	0.23	0.10	64.3	0.21
B_{21}	12-39	14.9	20.5	64.6	5.8	0.05	–	11.64	4.04	3.04	0.27	0.13	64.3	0.20
B_{22t}	39-57	23.1	20.3	55.6	5.6	0.04	–	13.63	3.56	3.05	0.27	0.16	52.0	0.18
B_{23t}	57-90	25.1	21.4	53.2	5.3	0.04	–	13.03	4.58	2.03	0.34	0.13	54.3	0.19
B_{31}	90-123	15.6	23.4	61.0	5.2	0.05	–	11.42	4.61	1.53	0.22	0.19	57.0	0.17
B_{32}	123-150	13.1	24.4	62.5	5.6	0.04	–	11.02	4.61	1.02	0.22	0.16	54.5	0.15
Erumaikulam														
A_p	0-20	13.1	13.6	73.3	6.9	0.06	–	14.12	6.10	4.60	0.22	0.22	78.9	0.22
A_{12}	20-42	15.8	14.9	69.3	7.3	0.09	–	13.02	5.09	3.56	0.27	0.26	70.5	0.19
A_{13}	42.54	12.9	15.6	71.5	6.3	0.08	–	13.12	5.60	2.55	0.25	0.33	66.5	0.17
Perumalpatti														
A_p	0-10	13.6	10.9	75.5	6.4	0.02	–	13.51	5.57	1.52	0.16	0.16	54.8	0.20
B_{21t}	10-32	14.6	17.4	63.0	5.5	0.04	–	14.82	4.60	1.52	1.52	0.13	57.2	0.23
B_{22t}	32-54	22.3	19.1	58.6	5.5	0.04	–	14.24	7.38	0.52	0.52	0.14	58.1	0.24
B_{23t}	54-70	17.5	22.0	60.5	5.4	0.07	–	14.00	7.18	0.04	0.04	0.13	54.1	0.20
B_3	70-90	15.7	22.0	62.3	5.6	0.04	–	13.00	6.85	0.13	0.13	0.20	57.3	0.21
Ramanujapuram														
A_{1p}	0-15	44.0	14.6	41.4	8.0	0.11	10.0	40.20	18.20	16.05	1.66	0.58	90.7	0.20
A_{12k}	15-42	52.3	21.6	26.1	8.0	0.09	9.5	43.01	21.61	15.79	1.77	0.48	92.1	0.21
A_{13kss}	42-65	52.6	21.4	26.0	7.9	0.21	9.0	45.70	21.01	19.01	1.63	0.38	81.5	0.23
A_{14kss}	65-110	54.6	23.1	22.3	8.0	0.58	11.0	44.93	18.76	17.08	1.83	0.58	85.4	0.20

Organic carbon is an important and discriminating diagnostic criterion. In general the soils were low in organic carbon content and it varied from 0.15 to 0.32 per cent in surface layers. The low values recorded clearly revealed the poor accumulation or organic matter under the prevailing environmental conditions.

SOIL CLASSIFICATION

The morphological, physical and physio-chemical properties were used for classifying the soils up to family level as per USDA Soil Taxonomy. The thirteen pedons were carefully studied for their morphological, physical and physico-chemical properties. Based on number, order and thickness of horizons, texture, structure, colour, organic carbon content and soil reactions, soils of each pedon were recognised as a distinct soil series. These soil series were classified as per USDA Soil Taxonomy at family level.

The soils of pedons Sevalpatti, Tiruchuli, Thimmapuram, Udayanendal, Palayampatti, Taraganendal, Pillayarkulam, Mandriodai and Kallatur were grouped under inceptisols. The presence of warmer isohyperthermic temperature regime and ustic moisture regime resulted in placement of these soils under the suborder Ustepts. The Sevelpatti soils series was grouped under Vertic Ustopepts at sub group level because of occurrence of deep, wide cracks within a depth of 50 cm during summer and slickensides within 100 cm of the surface.

The Tiruchuli, Thimmapuram, Udayanendal, Taraganendal and Mandriodai series were devoid of any diagnostic horizons and free $CaCO_3$. The base saturation was less than 60 per cent. The above factors resulted in grouping them under Haplustepts at great group level. These soils represent the central concept of the great group, accordingly named as Typic Haplustepts at sub group level. However at family level they were grouped into different families due to variations in textural and mineralogical classes.

The paralithic contact within 50 cm of the soil surface resulted in classifying the Palayampatti series under Paralithic Ustopepets at sub group level. The Pillayarkulam series was placed under Aquic Haplustepts due to the presence of mottles in the sub surface horizons within 100 cm from the mineral surface and aquic conditions for some time in a normal year. Lithic contact within 50 cm of the soil surface along with properties of great group Ustepts led to the Kallatur series to be placed under Lothic Ustepts. At family level it was classified as loamy skeletal, mixed, isohyperthermic, Lithic Ustepts.

The Avarangulam and Perumalpatti series were found to be Alfisols, because of presence of argillic horizon as evidenced by the accumulation of illuvial clay and iron oxides. The presence of Ustic moisture regime resulted in grouping them under Ustalfs sub order. The argillic horizons with a base saturation of less than 75 per cent throughout the pedon resulted in placing the Avarangulam series under Ultic Halplutalfs at sub group level. The Perumalpatti series was found to have an increase of more than 20 per cent clay within a vertical distance of less than 7.5 cm with BSP less than 75. Hence, it was grouped into Ultic Paleustalfs at sub group level.

The presence of ochric epipedon, ustic moisture regime, and clay content less than 18 per cent in particle control section resulted in placing the soils Erumaikulam series under coarse loamy, mixed, isohyperthermic, Typic Ustorthents at family level.

The Ramanujapuram series was classified into Typic haplusterts a sub group level because of presence of deep, wide cracks during summer and absence of any diagnostic horizon within 100 cm of the soil surface. At family level it was placed under fine, monmorillinitic, isohyperthermic, Typic Haplusterts.

Soil series or soil association in each mapping unit was identified. The soil boundaries were transferred to the base map of 1: 50,000 scale prepared from SOI toposheets. The soil map was given in Figure 12.2.

CONCLUSION

Remote sensing satellites provide a vintage point in space to get a synoptic view of the earth's surface in a repetitive manner. This makes it to monitor the environment and natural resources. The satellite imageries were extremely useful in identification and delineation of soil resources. The soils were widely varied in physical and physio-chemical parameters. The soil resource map thus generated can be used in development planning and technology transfer.

REFERENCES

Anonymous (1971), *Soil Survey Manual*, All India Soil and Landuse Organisation, Indian Agricultural Research Institute, New Delhi.

Buol, S.W., F.D. Hole and R.J. McCracken. (1980), *Soil Genesis and Classification.* The Iowa State University Press, Ames.

Challa, O. and S.T. Gaikewad, (1986), Soils of Catena from Dadra Nagar Haveli—Their Characterisation and Classification. *Journal of Indian Soc Soil Sci.* 34: 543-550.

Diwakar, D.P.S. and R.N. Singh (1994), Characterisation of Soils of Diara Land Occurring in Gangetic Plains of Bihar. *J. Indian Sco. Soil Sci.* 42(3): 420-425.

Jackson, M.L. (1973), *Soil Chemical Analysis*, Prentice-Hall of India Pvt. Ltd., New Delhi.

Menon, P.K.R. and A. Mariakulandai (1957), 'The soils of Madras', *Madras Agric. J.* 44: 313-325.

Soil Survey Staff, (1998), '*Keys to Soil Taxonomy*', Eight edition, U.S. Dept. Agril. Hand Book, 436, Washington, D.C.

SECTION IV

AGRICULTURE RELATED FIELDS

13

AGRICULTURE LIMITS AND ALTERNATIVES TO ITS GROWTH BASE

NANDESHWAR SHARMA

BACKGROUND BASICS

Agriculture forms the most fundamental life sustaining eco-system. Though the most common world's cultural phenomenon, it inherits necessary spatial manifestation in being never uniform or alike in either extent or kind. Agriculture signifies essential *human exercises to practise manipulation and management of land resources to ensure subsistence and sustenance of mankind.* It has represented, in sequences, to be the natural consequence of development – a growth process outcome in itself. As a growth outcome, it has meant the most natural, frequent in occurrences, varies and *fragile spatial manifestation of the first apparent cultural revolution* and has progressed to acquire generative and accumulative sequences of growth through time and areas.

Agriculture happens, in its present form, to be both resource and resource based pursuit. It is, undoubtedly, the first organised cultural artifact that has been producing and supplying resource inputs to support most economic activities in secondary and tertiary sectors. It finds, on the other hand, its inherent growth base in the availability of land resources in its quantitative and qualitative dimensions. These aspects have combined to shape the ecogenic and technogenic faces of agriculture. The ecogenic ingredients (mode of occurrences of resource elements) largely set limits and alternatives to its growth base. The *technogenic* elements (the nature and the magnitude of use of resource elements) have, too, decided the kind and the extend of *limits and alternatives* to the base and the stage of the growth of agriculture world wide. Set in to operate within ecogenic ingredients and man-induced technogenic elements, agriculture has continued to grow through *change-synergy nexus mechanism* which seems to be the condition in centre to determine its limits and alternatives to growth base.

Agriculture has responded to primordial man – food equation experiencing a balancing support of technology. The basics to growth of agriculture lie in the need to reorganize it and to effect land resources management. This necessitates the manipulation and management of limits and alternative to which agricultural pursuits are put in. The humanity has to aim high at ensuring a condition of growth of agriculture at the level of sustainability. The background basics of operational mechanism of agriculture has to be comprehended through the model outlined below:

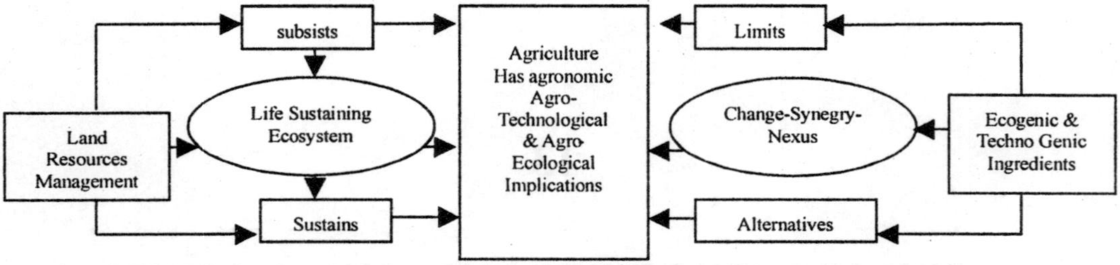

Agriculture has not been some object or pursuit of sudden or temporary appearance. It marks the part of a story of cultural civilization – the story of evolution, the story of change, the story of continuance and the story of march with human civilization.

THE STORY THAT BEGAN

The story of agricultural growth has been, inherently, composed through successive events of man's efforts to search for food in his environmental habitat. Man, food and environment have been the most fundamental components of the story that inscribes the emergence and evolution of human civilization. In his natural habitat, man wandered, gathered and collected naturally gifted articles for food. He, later, toiled to hunt animals to consume as foods. Though largely a searcher for food, he was constantly threatened to face starvation.

The vent of man's transition from hunter to tiller tended to write a new story of remarkable change in man's relationship with his environment. The change came ten to fifteen thousand years ago when man began to husband animals and plants. His intervention with the environment became apparent to manifest man's capacity to act to meet his needs for food. He found some selected conducive environment to experiment with his new pursuits in search of food. The fertile crescent of the Middle East proved to be, perhaps, the first theatre where drama of cultivation was enacted to procure and taste food crops. The spread of agricultural pursuits ensured increased stability of food supply. With it an increase in population, too, was witnessed. Of the two, which one became the cause and effect of each could not be clear (Evans, 1980) One thing, however, became clear that man, as a tiller, developed a seemingly endless prowess for altering his environment (Brown and Eckholm, 1975).

The story of the emergence and continuance of agriculture appeared to be composed in two distinct parts—(i) Palaeolithic Performances and (ii) Neolthic Accomplishments. The former signified the traditionality while the latter qualifies the modernity. Palaeolithic achievements assisted mankind in securing only a subsistence base. The Neolithic achievements have ensured reliability and hope to procure in abundance to feed and support increasing masses and economic pursuits of commercialization and industrialization in present day civilization where we are witnessing the phenomenon of increasing urban and diminishing rural.

THE STORY THAT CONTINUES

Given a beginning, agriculture has continued to grow and has grown to continue. The history of agriculture suggests that men set the initial trend of food crop production, continued to intensify its operational pursuits and continue to transform the environment with the support of ever evolving technologies at hand. The present day agriculture continues to necessitate the comprehension of multiple issues of *agronomic, agro-technological and agro-ecological* implications. It conforms an ecosystem within a *cultural triangle* with three apices of agronomy-technology and agro-ecology.

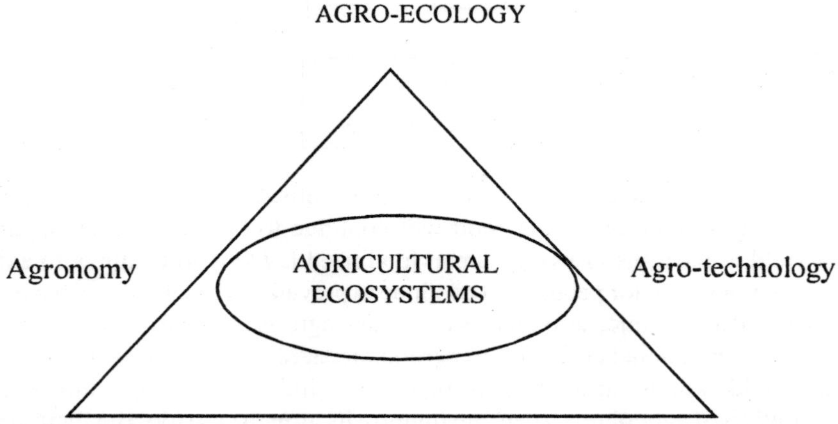

The agronomic comprehension of agriculture corresponds to the knowledge of the theory and practice of field crop production and the agricultural resource management. Agronomy, as a component of agricultural ecosystem, means simultaneously the science of food production and that of education to the public to create better appreciation of farming.

The agro-technological comprehension of agriculture relates to the condition of the assemblage of tools, techniques and skills to manifest the specific applicability to the culture of productivity of land. Agro-technology continues to set structural and operational equilibrium of agricultural ecosystem. It is aimed at assuring manipulative and managerial efficiency of land resources and agricultural resources to achieve sustainable ground for development.

The agro-ecological comprehension of agriculture has generated pertinent debate on issues of techno-added agronomic transformation as against impending agro-ecological crises. It senses the change in the scope of sophistication of agro-ecology in the wake of ever expanding manipulation of land resources by man—the "ecological dominant". Agro-ecology, thus, carries meaning to: (a) geographical appraisal of agricultural activities environmental stress, (b) assessment of nature and extent of agricultural resources base causes by human manipulation and management and (c) changes in the structure, functioning and equilibrium of agricultural ecosystem – largely an artificial ecosystem.

The story of world agriculture continues to (i) carry the past performances, (ii) sustain the present accomplishment and (iii) prepare to face the future. The nature and the level of agriculture are essentially accredition of agronomic, agro-technological and agro-ecological accrual through time and space. The form and the function of agricultural ecosystem may be sequenced to give the following stage of progress:

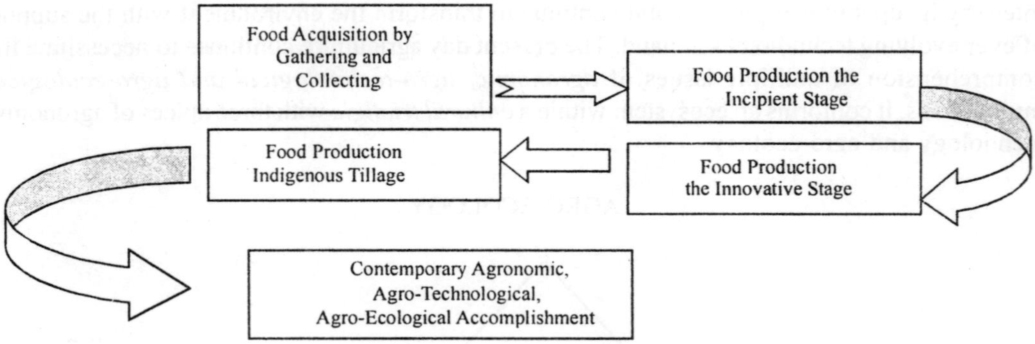

The story does not, however, end here. Given the situation, where world population will continue to increase, more and more food will continue to be required, an apparent lag in agricultural yield to meet the growing human food need is likely to continue to exist and the demand for better nutrition for people will continue to prevail, the mankind will have to continue with the practice of agronomic, agro-technological and agro-ecological change – synergy nexus. It is, however, the operational efficiency of agronomic, agro-technological and agro-ecological elements that holds key to structural and functional equilibrium in agricultural ecosystem. There exist frontiers of agriculture. There are limits to its growth base too. But there are instances that frontiers have been won, limits have been scaled and alternatives have come to rescue the humanity from deluge. Such instances have to be repeated to bridge the high gap between potential yield and actual yield of various crops" (Singh, 2000). Frontiers and limits do assert to continue and to force agriculture to mark uneven development, "characteristic of the so-called development as well as developing world" (Rice, 1999).

FRONTIERS OF AGRICULTURE

Agriculture represents the most common culturo-spatial phenomenon. It appears, however, certainly not the phenomenon of universal occurrences. It does have frontiers in geographic scales. The physical spread of agriculture follows the basic principles of economic development i.e. it occupies sites that are most acceptable, most accessible and can be developed with minimum cost and at least effort beyond the existing periphery. The degree of accessibility and acceptability of sites varies in geographic scales. Farmers have, accordingly, developed systems of agriculture that are subjected to modifications in time and area scales. This has helped them to define "agricultural realms or domains" (Singh and Dhillon, 1994) over the earth's surface. The world's agricultural realms extend as far as they meet frontiers, limits to prevalent growth alternatives.

The notion of frontiers of agriculture is conceived, at its first appearance, in recognition of two worlds – *ecumene and non-ecumene*. The physical limit of *non-ecumene* world maps the area extent of frontier of agriculture. It is believed that "half of the land surface of the earth is desert or semi-desert" (Pearos, 1967). This attests the "greatest sensitivities to climatic elements" (Haggett, 1979) that agriculture tends to have. The vast marine segment of the globe forms another apparent frontier of agriculture. Even in parts of *ecumene* world, there are sections of land, big or small, that make frontiers of agriculture in the form of wastelands, wetlands and other uncultivable lands. Frontiers with reference to initial phases of development of agriculture had prevailed over the cultural practices, the number of which "remaining for expanding the quality of land under cultivation has been found to dwindle" (Luoma, 1984) as relevant to modern agriculture. Frontier has always tended and continues to exist in varying quantitative and qualitative dimension. The dimensions of frontier have been changing in response to nature and degree of interactions among limiting components of natural environment on one hand and technological components of human environment on the other. Terranean, sub-terranean, edaphic and climatic resistance act to form *barriers to expansion* of agricultural activities which, in turn, demarcate the geographic scales of frontier of agriculture. Mankind responds, through innovation and application of technologies, to remove resistances of limiting factors and to capture the frontiers in succession. This leads to shrinking of frontiers in the wake of *carriers of alternatives.*

BARRIERS OF AGRICULTURAL DEVELOPMENT (Limits to Growth Base)

Agriculture appears to manifest a developmental consequence. It signifies, in this sense, a necessary growth process outcome. Agricultural development has also been proving a continuous source of fulfilment and frustration to those who practice them. This is so because the practice of agriculture functions under set growth condition, which forms its growth base. World experiences tell that the prevalent growth base does not always respond to favour, but confronts agro-ecological, agro-technological and agronomic deterrents too. This sets limits to the growth base of agricultural development. The nature and the extent of limits to growth of agricultural development become the function of interplay between critical components and necessary resistances that tell upon the operational mechanism of agriculture.

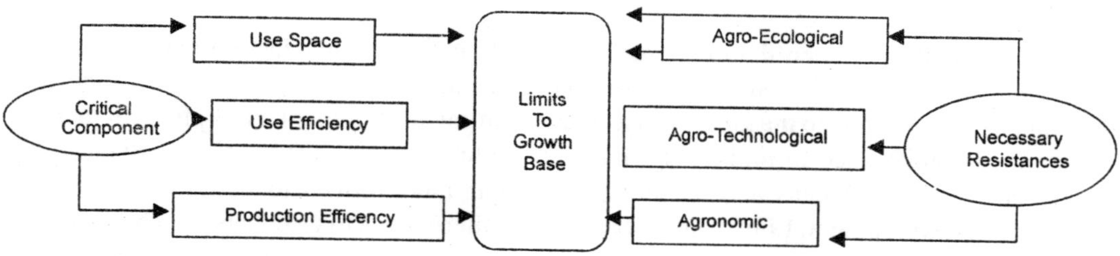

The instance of barriers acting upon the success of agriculture substantiates the observation that it progresses under unavoidable interaction among "biotic potential (productive potential), environmental resistances and carrying capacity" (Haggett, 1979). Agricultural scenario, as developed in its contemporary form world over, has amply shown that under the available technologies, groups of man kind have viewed the environmental base of agriculture in different ways in tempo-spatial perspectives. Agricultural development across the earth's surface relates to cultural and environmental factors. Elements of physical environment, viz. climate, soil and surface are found to "set broad limits on agricultural practices" (Rubenstein, 1989). Elements of cultural environment, viz. customary beliefs, performances, technological capabilities, social organisation and some others, too, decide the effectiveness of limits on agriculture. World geography has exhibited pronounced barriers in true spatial context, and so do agricultural pursuits. This has destined agriculture to inherit barriered and checkered limits in spatial and efficiency terms. Some of apparent limits, that can be identified, are as listed below:

- Expanse *non-ecumene* frontier in the presence of land segments facing terranean and climatic deterrents, high mountains, deserts-hot and cold, and bad land topographic highlands, for examples.
- Natural hazards susceptible frontier of coastal and inland locations, cyclone-torn coasts of Orissa, Andhra Pradesh, Tamil Nadu, Kerala and Kutch; water deficient, flood-prone, drought – prone and desert dominant land located segments, for examples.
- The emerging situation of intense human concern to compete seriously for space and survival in the wake of unrestricted growth of population causing ever diminishing share of land per man or shrinking use space for agriculture.
- Lack of technical know-how among farmers and inadequate and improper development and application of agro-technology on parts of practitioners of agriculture.
- Decreasing efficiency caused by loss of soil fertility and diminishing use efficiency engineered by the urge to procure maximum through wanton use of new agricultural technology.
- Deficient knowledge of regional capacity with reference to quality of land, local species specifically adapted to the climate soil and biotic factors and use of suitable technology leading to agro-ecological disequilibrium.
- Urban invasion and succession: non-agricultural uses are eating land base for agriculture fast.
- Poor agricultural education and information system restricting dissemination and perfection of agricultural practices.

There have always been determinants of agricultural operation. Determinants, both of *physical and non-physical* groups have contributed positively and negatively to decide the nature and extent of limits to the growth base of agricultural development. Non-physical factors have attained significance to mollify the intensity of limits by modifying the resistances of physical environment. Means and technological inputs have proved to function as carriers of agricultural development. Limits have been won by alternatives to growth base. The sense of need has initiated agriculture to stay and the practice of alternatives has set agriculture to continue and to win the frontiers.

CARRIERS OF AGRICULTURAL DEVELOPMENT (Alternatives to Growth Base)

The world agricultural scene displays obvious contrasts in its operational pursuits and fundamental differences are reflected sharply with reference to relatively rich and poor nations. Alternative forms of agricultural system have progressed in necessary association with the prevailing resource base and technological improvement, grounding the regional capacity. The world is witnessing period of marked rapid change and agriculture, too, has responded to progress in rapid change situation. "Highly mechanised 'AGRIBUSINESS' in the U.S. contrasts with the basic subsistence form common in many other parts of the world" (Hardwick & Holtgrieve, 1990). The emergence of agriculture has definite significance to mark the earth's most radical revolutions in human cultural and economic development. Contemporary apparently complex patterns of agricultural system produces marked variations in types of land resource management efficiency in direct response to opportunities available in terms of the quantity and quality of physical and cultural inputs. China's labour intensive rice cultivation has its contrasting alternative in Central Valley of California, which excels in rice production on a massive scale with the application of modern farm technologies on relatively large area scale.

A systematic and specific evaluation of agricultural land resources management indicates that managers practise alternatives. Alternatives are decided as per the nature and degree of interactions among physical, socio-economic and technological determinants at regional scales. Decision as to adopt alternatives at any particular time determines the "efficiency and effective use of agriculture and resource" (Pragathi and Ramanaiah, 1999), which ensures "the degree of optimal use and performance of cultivated as well as cultivable land" (Reddy and Ramanaiah, 1985).

Mankind has been practising alternatives to harness the agricultural growth base continuously to overcome the necessity of: (i) more food for ever growing human number and (ii) better nutrition for people by applying quantitative and qualitative improvements in land resources management. Among the most pressing needs of agricultural alternatives are: (i) procuring high production efficiency, (ii) enhancing efficiency in use of space available for agriculture and (iii) enlarging the physical space, i.e. taming the frontiers of agriculture. The situation in the context of Indian agriculture has indicated that (a) the new technology brought about "high growth rate" and (b) the geographical land area available for agricultural expansion, per capita "has been declining fast", (c) modernisation of agriculture has been restricted because of widespread poverty and "traditional structure of land and production" (Pathak, 2002).

The economic structure and the spatial processes of agriculture have shown that the preceding century has brought unprecedented agro-technological and agronomic transformation. The revolutionary transformation has made agriculture to act farming, business and research. Search for alternatives has become the sole fundamental alternative to strengthening the growth base to make agriculture sustainable. Sustainable agricultural development has to be treated as an alternative to form the part of the most essential human agenda. Alternatives to translate the agenda may include:

- *Use pace building:* It is possible through making unused territorial segments available for agriculture and reclaiming the cultivable wastelands. The scope lies in abundance.
- *Use efficiency building:* It is possible through effecting land rapprochement i.e. ensuring the parts of land to use best suited to.
- *Production efficiency building:* It is possible through the application of agricultural technologies at optimum level, i.e. in conformity with agro-ecological base of a particular region.
- *Change-synergy nexus building*: Change is not abstract and limited. It produces related effects. Any alternative course of action has to assure allied or related sectors protection against disequilibrium as changes in the level of one component cause changes in many other relevant components.
- *Making agro-education a people's movement*: Education enhances practitioner's receptiveness to new ideas and capacity to judge and practise the right alternatives.
- *Transformation of agricultural practices and systems* in the light of continuing processes of change from rural to urban and from agri to industrial.

- *Viewing agriculture not just as a sector of producing crops,* but treating and practising it as part of comprehensive land resources management emerges to be a necessary alternative.

An approach to comprehensive land resources management is understood to be key to achieving stable agriculture to sustain. A balanced spatial structure, effective and best-suited agro-technology, organisational understanding and skill of agronomy and change-synergy-nexus can ensure an acceptable and resources management and thus a sustainable agriculture. The resource base is, however, not uniform at all places and it tends to support agriculture and for the matter economic system differently.

SOME IMPERATIVENESS

Growth base of agriculture has been both inherited and created. Mankind has been, accordingly, manipulating and managing agriculture by applying his might and skill. This character of agricultural growth base to exist and assert between two prongs of environmental dictates and human responses has made agriculture an issue of *behavioural and applied implications.* Changes has been the most pronounced behaviour and adoption and adjustment by men to new changes warrants application of knowledge and management skills. A systematic information network is considered vital to the comprehension of problems and prospects of land use planning which, in turn, becomes necessary for making agriculture sustainable. A comprehensive inventory of facts and figures, a detailed map of land resources available and patterns of land use developed are two priori of agricultural growth base and related alternatives. The development and availability of Geographic Information System (GIS) and the application of remote sensing techniques to the preparation and interpretation of land use maps have attained increased importance. And this is not without reason. It is opined – "Land use data are needed in the analysis of environmental processes and problems that must be understood if living conditions and standards are to be improved or maintained at current levels" (Anderson, *et al.*, 1971). It is also observed that 'rapidly developing technology of remote sensing offers an efficient and timely approach to the mapping and collection of basic land use and cover data over large area" (Gautam, 1982). Adequate information system developed in connection with agro-ecological agro-technological and agronomic base of growth can ensure the effectiveness of alternatives to achieve the development of agriculture at sustainable level. This theme may be comprehended through the following model:

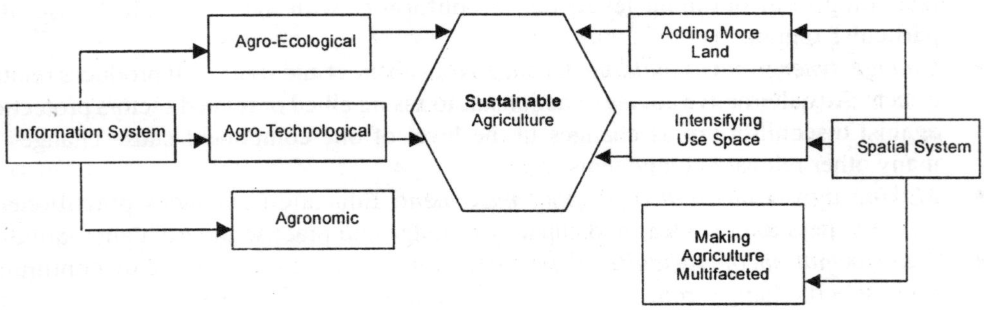

REFERENCES

Anderson, J.R., *et. al.*, (1971), A Land Use and Land Cover Classification System for Use with Remote Sensing Data, Professional Paper, 964, *U.S.G.C,* Reston, Virginia, p.

Brown, L.R. and Eckholm, E.P., (1975), Man, Food and Environment, *in Environment*, (ed) W.W. Murdoch, Sinauer Associates, Inc. Publishers Sunderland, Mass, p. 68.

Evans, I.T., (1980), The Natural History of Crop Yields, *American Scientist*, 68, p. 388.

Gautam, N.C., (1982), Suggested Land Use/Land Cover Classification System for India Using Remote Sensing Techniques, Pink Publishing House, Mathura, p. 1.

Haggatt, P., (1979), *Geographic: A Modern Synthesis,* Harper & Row Publishers, London, pp. 47, 49.

Hardwick, S.W. and Holtgrieve, D.G., (1990), *Pattern on Our Planet: Concepts and Theories in Geography,* Merrill Publishing Co., New York, p. 219.

Luoma, S.N., (1984), *Introduction to Environmental Isuses*, Collier Macmillan Publishers, London, p. 359.

Pathak, C.R., (2002), *Spatial Structure and Processes of Development in India,* Regional Science Association, India Kolkata, p. 52.

Pearos, L.C., (1967), *Principles of Agronomy,* Affiliated E.W. Press Pvt. Ltd., New Delhi, p. 12.

Pragathi, P.S. and Ramanaiah, Y.V., (1999), Regionalisation of Agricultural Land Use Efficiency in Andhra Pradesh, India, *IJRS,* Vol. XXXI, No. 2, *Regional Science Association*, Kolkata, p. 61.

Reddy, N.B.K. and Ramainaih, Y.V., (1985), Changes in Agricultural Land Use Efficiency in Andhra Pradesh, *Land Use Policy*, Vol. 2, No. 3, p. 211.

Rice, M.J., (1990), Uneven Economic Development: Examples from India and Georjia, U.S.A. *Annals of NAGI*, Vol. XIX, Nos. 1 & 2, New Delhi, p. 48.

Rubeinstein, J.M., (1989), *The Cultural Landscape: An Introduction to Human Geography,* Merrill Publishing Co., London, p. 309.

Singh, G.B., (2000), Green Revolution in India: Gains and Pains, *Annals of NAGI*, Vol. XX (No. 1), New Delhi, p. 18.

Singh, J. and Dhillon S.S., (1994), *Agricultural Geography*, Tata McGraw—Hill Publishing Co. Ltd., New Delhi, p. 144.

14

CLIMATE CHANGE AND AGRICULTURE: SOME EXAMPLES FROM KASHMIR

RAIS AKHTAR

INTRODUCTION

The natural and cultivated plants of mountain ecosystems may be affected by climate change in two different ways: first, by the increased concentrations of CO_2 available for photosynthesis (i.e. physiological responses); and, second, by the actual changes in climate (i.e. physical responses). These two types of responses are very closely interlinked, since physiological processes such as photosynthesis and respiration are directly influenced by temperature, water availability and, in the case of photosynthesis, sunlight.

Most assessments of the potential impacts of climate change have focused on increasing temperature, for the following reasons. First, this is the most obvious change, which would lead to other changes (e.g. soil moisture, evapotranspiration, fire frequency and severity). Second, it is the change which can be most easily assessed for past (paleo-) and recent climates, from which parallels for future climates can be drawn. Third, a large literature exists on the influences of temperature on the distribution of species and on ecological processes. The third aspect is important as it applies to water regimes.

However, there has been relatively little research on the potential impacts of climate change on mountain agriculture. In the context of natural ecosystems predictions have been made that climate change will lead to higher climatic limits to cultivation (Price and Barry, 1997).

From different climate change models it is predicted that the most affected regions and societies of the world are coastal and mountain societies. Reasons being the critical climate region they live in where a little change can bring major shifts (TIME, 1999).

CLIMATE CHANGE AND WORLD FOOD SUPPLY

Concern about world food production has increased since the mid 1980's when world per capital production for several major food items, especially cereal grains, appeared to reach a plateau (McMichael, 1996). Long term changes in global and regional climate would affect some aspects of agriculture. According to McMichael, climate change could influence food production adversely via:

(i) Geographic shifts and yield changes in agriculture;
(ii) Reduction in the quantity of water available for irrigation;
(iii) Loss of land through sea level rise and associated salinization.

However, higher temperature, more efficient use of water by plants might increase yields in some areas. But in regions where the available food is just sufficient for the populations, even the slightest decline in yields could be very harmful. For instance, climate-related declines in fish catches could pose particular nutritional stresses on human populations in less developed countries, for whom a high proportion of dietary protein is derived from seafood.

There is no denying the fact that apart from climatic factors, productivity trends in agricultural sector are strongly influenced by changes in national and international market forces, government policies, trade agreements, the development of new strains of crops and the availability of new technologies. Thus, as stated by the FAO the impacts of global environmental changes on agriculture would be greatly modified by the agriculture sector's own internal dynamics.

TEMPERATURE, PRECIPITATION AND CARBON DIOXIDE LEVELS

Increased levels of atmospheric CO_2 are predicted to cause global warming, which in turn, would alter regional surface temperatures and affect crop growth. It is already known, for example, that the accelerated maturation of wheat under higher temperatures reduces the growing time for wheat grains-and, therefore, their final size. Australian scientists simulated the combined effect of a 3°C increase in temperature, increased CO_2 (to 560 ppmv by 2050), and altered variability in precipitation and soil moisture. They concluded that although the yield of late – maturing strains would be increased, the yield of some early-maturing strains would be halved (McMichael, 1996).

Crop yields are also sensitive to changes in soil moisture. Several climate models predict a global increase in precipitation (mostly as greater rainfall intensity – i.e. more rain per rain day). There is likelihood that equatorial monsoon rains may extend to higher latitude regions in Africa's Sahel and in north-west India, and alleviate some drought conditions. Intensified precipitation would exacerbate flooding and soil erosion, resulting in destruction of crops and loss of arable land. It has also been predicted that precipitation may be reduced in some tropical and mid-continental regions. Seasonal shifts in annual precipitation patterns could also occur, affecting soil moisture during the growing season, and possibly impairing the growth of many crops. It has been argued that the inherent limitations of plant physiology render long-term adaptations to a continuing change in climate unlikely (Woodward, 1987, Quoted from McMichael, 1996). Food and Agricultural Organization is deeply concerned with the issues of global warming, climate change and variability. Climate change over the long term, in particular global warming, could affect agriculture in a number of ways—the majority of which would threaten food security for the world's most vulnerable people:

(i) The overall predictability of weather and climate would decrease, making planning of farm operations more difficult.
(ii) Climate variability might increase, putting additional stress on fragile farming systems.

(iii) Climate extremes – which are almost impossible to plan – might become more frequent.
(iv) The sea-level would rise, threatening valuable coastal agricultural land, particularly in low-lying small islands.
(v) Biological diversity would be reduced in some of the world's most fragile environments, such as mangroves and tropical forests.
(vi) Climatic and agro-ecological zones would shift, forcing farmers to adapt, as well as threatening natural vegetation and fauna.
(vii) The current imbalance of food production between cool and temperature regions and tropical and sub-tropical regions could worsen.
(viii) Pests and vector-borne diseases would spread into areas where they were previously unknown.

CLIMATE VARIABILITY IS FARMER'S GREATEST CHALLENGE TODAY

The natural variability of rainfall, temperature and other conditions are the main factors behind variability in agricultural production, which in turn is one of the main factors behind food insecurity. Some areas of the world are particularly prone to such variability – the Sahel, northeast Brazil, Central Asia, and Mexico, for instance.

According to FAO, climate extremes, violent and unusual events such as floods, droughts, and storms—though by nature more apparently dramatic, have less overall effect on agricultural production than chronic deficiencies. Both climate variability and climate extremes may increase as a result of global warming. The FAO further adds that agriculture is not only a victim of global warming.

At present, it is also a contributory factor, and in the future, it could make a major contribution to reduction of global climate change. Some 25 per cent of carbon dioxide emissions come from landuse change (mainly deforestation in the tropics), and fertilizer use is one of the main human sources of nitrous oxides. The FAO encourages farmers to reduce excessive use of nitrogenous fertilizers-(which are costly, inefficient and harmful to the environment) and to choose livestock breeds that are efficient converters of feed into meet and milk, thus reducing methane emissions.

KASHMIR SCENARIO

The Kashmir region, because of its unique orographic position, has very varied climate. The most discernable impact of the climate change has been felt is the change of seasons in Kashmir. According to Soz, "The entire pattern of seasons has undergone change... the peculiar to the valley, the TSONTH—a situation that accrued for centuries between the end of snowy winter and the spring. The TSONTH was essentially a rainy season that started roughly from first of March and closed in the first week of April ... that traditional winter (December—February) with heaps of snow and icicles hanging from the long rim of the roofs is gone. For two decades or more, snow has not been experienced in Kashmir in the usual manner" (Soz, 2000). July is the hotest month in Kashmir. The analysis of temperature data reveals that both maximum and minimum temperatures of Srinagar (5440 feet) in the Kashmir valley have risen in comparison to the foothill region of Jammu (1200 feet) and Delhi (6600 feet). This is the first time that such increase in temperature was recorded in the Kashmir valley.

The analysis of data maintained at the Meteorological Observatory of the S.K. Agricultural University in Srinagar reveals that there has been a substantial decrease in the amount of precipitation for last four consecutive years (1997, 1998, 1999, 2000). As a result, the valley has experienced two droughts continuously during 1999 and 2000. As against the long term average of 816.3 mm, the precipitation received was only 716.4, 680.5, 568.5 and 433.7 mm which was less by 99.9, 135.9, 247.8 and 382.6mm or 12.2, 16.6, 30.4 and 46.9 per cent during 1997, 1998, 1999 and 2000 respectively. The drought occurred in 2001, because the months of RABI sowing (Sept.-Nov., 2000) received only 40.1 mm precipitation as against cummulative normal of 97.1 mm. Interestingly, the month of October (2000), in which most of the crops are sown, ran completely dry against the normal precipitation of 39.1 mm. The farmers have either deferred the sowing of their RABI crops or sowing was done in anticipation of the rains. The month of October also recorded the higher temperatures of 25.6°C as against 20.5°C which was 5.1°C higher, further, contributing to evapo-transpiration.

During 1999-2000 agriculture sector has suffered a whooping loss of Rs. 355.6 crores following the severe drought and dry spell conditions. Official figures suggest that around 1.75 lakh hectares of land did not yield any crop following the drought. According to the Directorate of Agriculture, "Fifty-one per cent of irrigated land suffered due to the dry spell and major crops like maize and paddy suffered badly… nearly 51 per cent of paddy crop failed due to drought. Besides, 69 per cent of maize, 55 per cent of pulse and 50 per cent of fodder crops failed in 1999-2000.

Some of the recommendations made to combat such situation are:

(1) Manual weeding to be adopted instead of hoeing, so that moisture is retained in the soil.
(2) Additional top dose of nitrogen to be added to mustard crop, in order to prolong the vegetative phase, which is otherwise, likely to suffer damage at the following stage.
(3) Choice of crops should be made purely on the moisture availability and their ability to withstand moisture stress.
(4) Surplus water in the early stages of availability need to be conserved for use as life saving irrigation.
(5) Short duration and less water intensive crops should be preferred to long duration and more water intensive crops.
(6) Hybrid seeds should be avoided.

CONCLUSION

Agriculture is the backbone of mountain societies. As opined by the IPCC in developing countries the mountain communities would suffer a lot if appropriate measure are not adopted. The management of water resources is the key to prosperity. A decline in rainfall and rise in temperature would lead to low soil moisture over the region and reduced water availability for agriculture and other domestic purposes. Confident predictions regarding future changes in the monsoon pattern due to expected climate change are needed to assess its impact on agriculture, hydrological resources and biota. According to UNESCO document, these predictions can enable the policy makers to foresee future problems, initiate appropriate response strategies to minimize their impacts.

REFERENCES

McMichael, A.J. and Kovats, (1996), *S. Climate Change and Human Health*, WHO, Geneva.

Price, M.F. and Bary, R.G., (1997), *Climate Change*, in B. Messerli and J.D. Ives (eds) Mountains of the World: A Global Priority, The Parthenon Publishing Group, New York, 409-445.

TIME, (1999), Newsletter, *Technological Intervention for Mountain Ecosystem*, Vol. 2. p. 5.

Soz, S., (2000), Climate change: The Kashmir experience, in S. Soz (ed.) *Climate Change Impact and Awareness,* People's Empowerment Mission, Srinagar, 113-118.

15

POPULATION AND FOOD RESOURCES IN SULTANPUR DISTRICT (U.P.)

DR. SATYAKAM ARYA

INTRODUCTION

A starving man thinks of satisfying his hunger before anything else. He will sell his liberty and all for the sake of getting a morsel of food. Such is the position of the millions of people in India. For them Liberty, God and all such words are mere letters put together without the slightest meaning. Today the shortage of foodgrains is one of the most burning problems before human beings in the world. The production of food grains is not growing so rapidly and in the same ratio as the growth of population. As a matter of fact, physical, economical and social status of man is deteriorating day by day. Indeed the problem of fulfilment of foodgrains is neutralizing all the planning for human development. Due to imbalance between growth of population and foodgrains production, this problem is still existing in developing countries. In this perspective Chaudhari (1952) has expressed his view that only 28 per cent area and 18 per cent population have extra foodgrains whereas 72 per cent area and 82 per cent population are always in shortage of foodgrains even for their own survival in India. The study area is also not exceptional and free from the problem of foodgrain shortage.

THE GEOGRAPHIC BASE OF THE AREA

Sultanpur is the second southern most district of Faizabad Division in U.P. It lies on the both sides of the river Gomati between the parallels of 25° 59′ and 26° 40′ north latitudes and 81° 31′ and 82° 41′ East longitudes. It is bounded by Faizabad on the north, Jaunpur and Azamgarh on the east, Pratapgarh on the south and Raibareli and Barabanki on the west and north east. It has a total area of 4,460 sq. km. It consists of 19 development blocks having 2488 inhabited rural settlements in which 2558970 persons reside (1991). The district is a part of the vast Gangetic alluvial plain. General slope of the land is from northwest to southeast. The drainage system of the area follows the general slope. On the basis of the topography, drainage and soil, the study area may be demarcated into Banger (Old alluvium) and Khadar (New alluvium). This region has an old history of intensive cultivation. Agriculture is the primary occupation of the people. The net area sown covers nearly 66.94 per cent of the total area.

METHODOLOGY

Per head availability of foodgrains is related to food production and population of a region. This is a ratio between food production and population. The annual availability of per capita foodgrains in the study area has been worked out at development block level in kilograms by multiplying the area with the yield of respective crops. As the requirement of food varies according to age, occupation, location etc., the Rusk's average coefficient which is also called as coefficient of consumption unit has been used. Foodgrains availability index for balanced diet and annual availability of foodgrains for per adult unit has been calculated by dividing the total available food with total consumption unit. The availability of foodgrains per adult unit is divided by the standard requirement of foodgrain for one adult unit which gives Foodgrains Availability Index. Foodgrains Availability Index is classified in to three categories i.e. low (below 148.92), medium (148.92-177.73) and high (above 177.73)

The formulae adopted for the calculation are as follows:

(i) Total adult unit = Population × 0.84 (Rusk's average Co-efficient)

(ii) $$\text{Annual availability of foodgrains} = \frac{\text{Total Annual foodgrains}}{\text{Total adults unit}}$$

(iii) $$\text{Foodgrains availability index} = \frac{\text{Per adult unit quantity of available Foodgrains}}{\text{Per adult unit required quantity of foodgrains for balanced diet}}$$

According to Nutrition Advisory Committee an adult needs 144.87 kg. Foodgrains as balanced diet.

Table 15.1: Relation Between Food Production and Population in District-1991

Development Block	*Total Foodgrains Production*	*Total Adult Units*	*Foodgrains Availability Per Adult unit*	*Foodgrains Availability Index*
1. Jagadishpur	260103	117898	220.61	152.28
2. Jamo	2454240	96161	255.03	176.04
3. Mushafirkhana	188464	89842	209.71	144.88
4. Baldirai	204390	83906	243.59	168.14
5. Amethi	164925	62026	265.89	183.53
6. Bhadar	181452	78687	230.59	159.17
7. Bhetua	221079	73878	299.28	206.53
8. Gauriganj	270390	30773	298.87	205.61
9. Kadipur	237653	10770	220.66	152.31
10. Lambhua	184965	83167	222.40	153.51
11. Pratap Pur Kamaicha	122981	60238	204.15	140.91
12. Dostpur	261376	90548	288.66	199.25
13. Akhand Nagar	253133	91708	276.02	190.52

(Contd.)

(Contd.)

Development Block	*Total Foodgrains Production*	*Total Adult Units*	*Foodgrains Availability Per Adult unit*	*Foodgrains Availability Index*
14. Jaisingpur	203035	90168	225.17	155.42
15. Dhanpatganj	179800	82111	218.85	151.06
16. Kudebhar	199698	90954	219.55	151.54
17. Kudwar	179462	93699	191.53	132.20
18. Dubepur	165020	94830	174.01	120.11
19. Bhadaiya	176528	81028	217.86	150.38
District	3898440	1715933	227.19	156.82

Table 15.2: Development Blocks According to Category of Foodgrains Availability Index

Category of Foodgrains Availability	*Range of Foodgrains Availability Index*	*Development Blocks*
High	Above 177.73	Amethi, Bhetua, Gauriganj Dostpur, Akhandnagar (5)
Medium	148.92-177.73	Jagdishpur, Jamo, Baldirai, Bhadar, Kadipur, Lambhua, Jaisinghpur, Dhanpatganj, Kurebhar, Bhadaiya (10)
Low	Below 148.92	Musafirkhana, Pratap Pur Kamaicha, Kudwar, Dubepur. (4)

RESULT

Table 15.1 shows that maximum availability of Foodgrains (above 177.37) is found in Amethi (14.33), Bhetua (206.55), Gauriganj (205.61), Dostpur (199.53) and Akhandnagar (190.52) whereas the required availability of foodgrains is 144.87 kg for balanced diet. All these blocks have more double cropped area due to adequate use of chemical fertilizers and available irrigation facilities by tube-wells. Second category region comprises the Jagadishpur Jamo, Baldirai, Bhadar, Kadipur, Lambhua, Jaisinghpur, Dhanpatganj, Kurebhar and Bhadaiya blocks where the availability of foodgrains index range from 148.93 to 1777.73. In this category by adopting good management practices relatively better crop production are being taken. Musafirkhana, Pratap Pur Kamaicha, Kurebhar, and Dostpur block belong to third category of region of low foodgrains availability where foodgrains availability index is below 148.92.

CONCLUSION

On the basis of this study it may be concluded that blocks falling under third category at low foodgrains availability region should pay attention first in optimum use of agricultural land. The main purpose of the study is to search the possibilities of maximum production of food-grains after outlining cultural factors which influence the better land use.

REFERENCES

1. Singh, S.P, (1991), Poverty Food and Nutrition in Indian.
2. Mohammed, A., (1978), Situation of Agricultural Food and Nutrition in Rural India.
3. Joshi, Y.G., (1972), Narmada Basin Ka Krishi Bhoogol.

16

SETTING AQUACULTURE FARMS ALONG THE COAST: A GIS APPROACH

T.A. RUBEENA AND C. CHINTH SHOBANA

INTRODUCTION

The coastal zone, the interface between the land and the sea, is an area of convergence of activities. The Indian Coastal Region has always remained attractive providing opportunities for the development of several industries like the thermal power, aquaculture, maritime activities, chemical and manufacturing, shipyards and tourism, to mention a few. For India, which possesses nearly 1.19 million hectares of brackish water area, coastal aquaculture is an important activity for employment generation and for generating the much needed foreign exchange. But the voice has been raised by environmentalists against the haphazard development of coastal aquaculture, polluting the coastal lands.

A very good example for the environmental deterioration, caused by aquaculture is that of the Radhanallur village of Nagapattinam district of Tamil Nadu. Here the drinking water has become highly saline, the agricultural lands have become unsuitable, and ponds have been polluted with chemical let outs from the aqua farms. Coastal zones have fragile ecosystems the survival of which depends on the right blend of coastal activities and their proper management.

To limit the level of environmental deterioration, special zones for aquaculture can be established on the basis of a set of selected criteria, after detailed investigations. In this way, proper development and management of farms, canals and services can be facilitated. Appropriate buffer zones can be created so that agricultural activities and freshwater sources do not suffer from salinisation. This is a tedious process, as demarcating the special zones manually is quite impossible (Alagarswami, 1993).

With the advent of the Geographical Information System, resource management has become an easier task. The data can be stored and is ready for immediate access and manipulation. Each variable (GIS layer) is achieved in a computer compatible digital format as a geographically referenced plane in the GIS database. The database may contain any type of information that is spatially distributed ranging from socio-economic to climatological to fundamental variables (Ramachandran, 1993; Krishnamurthy, 1993).

The present study tries to integrate the techniques of remote sensing and GIS to determine suitable sites for locating aquaculture farms, in Thirukalukundaram block of Kanchipuram district of Tamil Nadu. There are two basic objectives of this study: (1) To identify and evaluate each criteria for the analysis; and (2) Using the criteria to determine the sites, through the techniques of remote sensing and GIS.

STUDY AREA

The area chosen for study is the Thirukalukundram block of Kanchipuram district lying along the Northeast coast of Tamil Nadu. This area has been blessed with a number of creeks, inlets, estuaries, and mud flats, which are favourable for setting up aquaculture farms. It lies in between longitude 79° 57′ 10′′ East and 80° 12′ 10′′ East longitudes and 12° 28′ 30′′ North and 12° 41′ 30′′ North latitudes with an area of 352.82 square kilometres.

The Buckingham Canal flows from northeast to southeast. The predominant geomorphological features that are found in the block are buried pediments, pediment-inselbergs, residual hills, floodplains, sedimentary plains, beach and coastal plains. The hydrological soil group 'B' with moderate infiltration and moderate runoff potential covers 45 per cent of the area of the block. The soil group 'C' with slow infiltration and moderate runoff potential covers 29 per cent of the area of the block. The soil group 'A' with high infiltration and low runoff covers about 26 per cent while around 61 per cent of the block falls under wasteland, which is mainly barren land. Water bodies and agricultural lands occupy around 13 per cent and 11 per cent of the block, respectively, while forests cover 5 per cent of the block.

METHODOLOGY

The precise and logical steps that were followed to fulfil the objectives include field observations, semi-structured key-informant interviews, secondary data collection, criteria selection, preparation of thematic maps, reclassification and ranking of the thematic maps, followed by the performance of overlay analysis using GIS to determine suitable sites.

Field visits were undertaken to the aquaculture farms and the villages surrounding them in the Thirukalukundram Taluk in order to get first-hand information about the study area in terms of physiography and land uses. During the field visits, the villagers, the Panchayat leaders, members of the scientific community engaged in research here, the fisher folk, farm owners and farm workers were interviewed to collect their perceptions about the location of aquaculture farms.

The secondary data needed for the study were collected from relevant government and research organizations. Criteria for sitting were selected on the basis of information gathered through field observations, key informant interviews, and the Supreme Court Directive.

SELECTION OF CRITERIA

The criteria (Ramesh and Rajkumar, 1996) for the selection of sites can be broadly divided into two:

(1) *Regulatory principles set down by the Government*: According to the Supreme Court judgement dated 11th December 1996, the farm should be at least 300 m away from the seashore. The intention behind the judgement is to develop aquaculture in a sustainable and eco-friendly manner. The Government, for the use of land for aquaculture, has prohibited the use of agricultural and residential lands. Land within 1,000 m of fresh waterlakes, ponds, and rivers is not to be used for aquaculture, according to the Coastal Zone Regulations Act. This is so that even on a long run the water does not turn saline.

(2) *Geographical Criteria*: These are considered in locating an aquaculture farm. It includes quality of soil, drainage, coastal land uses, and transport.

(3) *Quality of Soil*: Soil's physical and chemical properties are important for efficient and sustainable practice of aquaculture. The soil should have an average pH of 7.5. Soil type and texture play an important role in retaining water. The bottom of the pond and sides should usually contain high clay content to prevent excessive seepage. Mostly, swamp soils contain peat. For this type of soil, the supply of oxygen is limited and peat does not decompose appreciably. If the soil contains pyrites and it is exposed during low tides, oxidation occurs. This is unsuitable for brackish water operations because this reduces the amount of oxygen in the soil, which indirectly affects the growth of prawns.

(4) *Drainage*: Sheltered bays, mud flats, lagoons, backwaters, creeks and estuaries are the best suited for aquaculture because they provide brackish water needed for aquaculture.

(5) *Coastal Land Uses*: Agricultural areas, forests, plantations, sandy and residential areas are least suitable; wetlands are found suitable while mud flats and tidal flats are most suitable for aquafarms because these are mostly low-lying, wasteland areas and the soils of these areas are also clayey. These characteristics make them suitable for setting up aquafarms.

(6) *Transport*: Close proximity to road will result in low road building cost compared to sites at distant places. Another important point is that prawns are highly perishable and have to be transported to the processing centres or markets as quickly as possible. Hence, it is best for these centres to be well-connected by road network.

MAPPING AND ANALYSIS OF THEMATIC MAPS

After the criteria were chosen they were mapped and subjected to overlay analysis using GIS to fulfil the second objective. The maps and images needed for the preparation of the thematic maps collected from the archives of the Institute of Remote Sensing, Anna University. Image interpretation and analysis was done to prepare the land use map, which is one of the most important maps needed for further analysis. IRS IC LISS III imageries taken on February 17, 1997 [path 102, Row 64 and path 102, Row 65] were used for obtaining the land use map. The interpretation key also helped further in identifying the land use zones based on zonal and textural variations. The task of preparing a land use map was simplified due to the availability of a well-defined reference map.

The hard copies of the maps were scanned and converted into digital formats using Auto CAD 2000. The individual themes were digitised as separate layers and were stored in drawing formats which were later converted into digital format. For editing the maps in digital format, they were transformed to arc format in ARC/INFO. Later the maps were edited for errors using Acredit options. Since Auto CAD provides options for snapping while digitising the errors could be minimised. The maps were then projected in ARC/INFO. Then the arcs were assigned unique ARC Ids and the unique polygons were identified within each map and were assigned unique label Ids. The label errors generated were rectified to get a flawless map. The unique polygons identified within each were assigned their respective names by preparing a polygon attribute table for the individual maps.

These maps were converted into shape (.shp) format in Arc View to prepare the layout for the maps. Legends, scale, and other needed information were incorporated into the maps. For transport as well as drainage maps, buffers were created separately using the buffer option in Arc View. The transport buffer as well as the drainage buffer maps were exported to ARC/INFO and the polygons were given unique label Ids and polygon attribute tables were made for both the maps.

RECLASSIFICATION AND RANKING

The different classes identified within each map were classified into three groups and ranks 5, 3 and 2 were assigned for the groups depending upon their suitability for locating an aquafarm. The weightages were then assigned to different criteria used in the overlay. Land use was given the highest weightage of 40. Soil was given a weightage of 25, geomorphology a weightage of 15, drainage a weightage of 10, transport a weightage of 6 and proximity to sea a weightage of 4. It is shown in Table 16.1.

Table 16.1: Criteria and Weightages Assigned

Criterion	*Rank*	*Features*	*Weights*	*Classes*
Land use	5	Mud flat, Wasteland, Inlet; Dryland, Sand,	40	High
	3	Plantation, Drainage, Cultivable land		
	2	Settlements, Tanks, Forests, Hills River Islands		
Geomorphology	5	Mud Flat, Buried Coastal Plain,	15	High
	3	Buried Deep Pediments		Moderate
	2	Buried Shallow Pediments		Low
Soil map	5	Clayey to clayey loam	25	High
	3	Loamy		Moderate
	2	Sandy		Low
Proximity to drainage	0	< 500 m	10	Low
	10	> 500 m		High
Proximity to Transportation	10	< 500 m	6	High
	0	> 500 m		Low
Proximity to Sea	10	> 300 m	4	High
	0	< 300m		Low

OVERLAY ANALYSIS

For the individual coverages, which were used for the overlay analysis, polygon attribute tables were created in ARC/INFO to assign ranks and weightages for the individual polygons. After assigning the ranks, the maps were overlaid, using the UNION command in ARCINFO. For the resultant map a polygon attribute table was prepared to represent the aggregate weights for all the polygons within the overlay. The map so obtained was converted into shp format in arc view and the legend was prepared for the coverage. In the coverage obtained potential areas for locating aquaculture farms could be demarcated as these were the areas with the highest aggregate rank. For each of the maps obtained, a layout was prepared. The criteria maps and the overlay were interpreted to obtain the results.

INTERPRETATION

The soil map is prepared on the basis of the United States Departments of Agriculture Classification. Of the soil groups, found in the region, Typic Ustipsamments-Typt, Udic Haplustalfs-Uhsf, and fluventic Ustochrepts-Fuop are the permeable sandy to silty soils, which are not suitable for locating aquaculture farms, since they allow the saline water to easily enter the ground water table. Udic Rhodustalfs-Ursf, Typic Ustochrepts—Tucp Typic Ustropepts-Tutp and Udic Ustochrepts-Uuop are the semi permeable silty to clayey loamsoils.

They are relatively better-suited for locating aquaculture farms. Entic Chromusters—Ecsv, Vertic Ustropepts-Vutp and Vertic Ustochrepts-Vuop are the most-suited owing to the impermeable nature of the soil. They are mostly clayey loams. (Fig. 16.1)

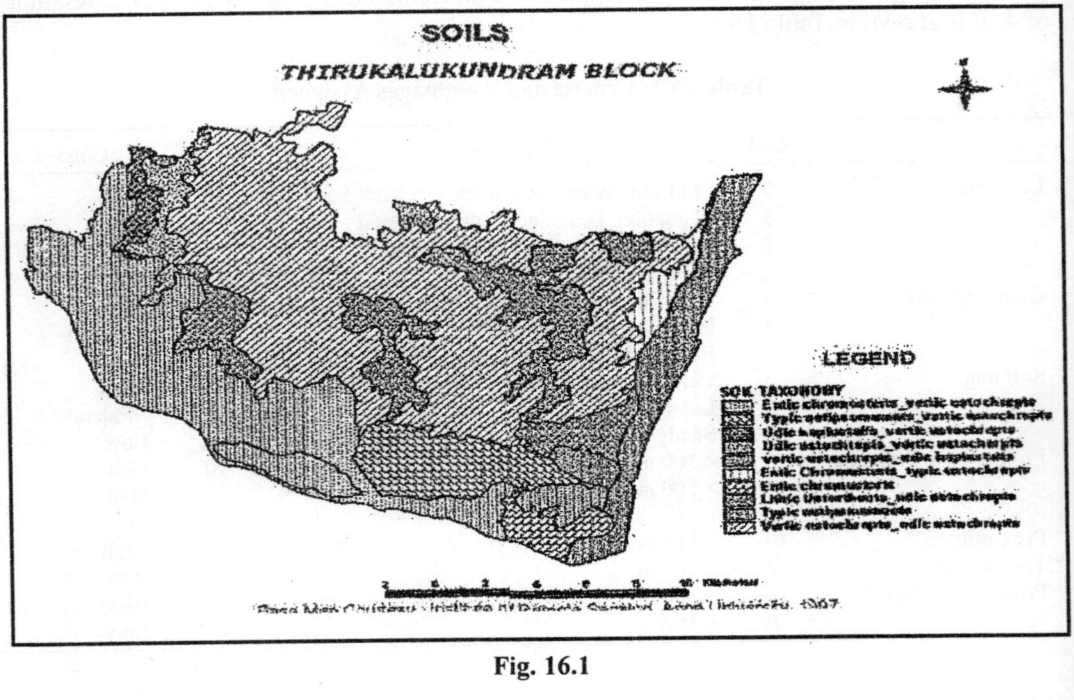

Fig. 16.1

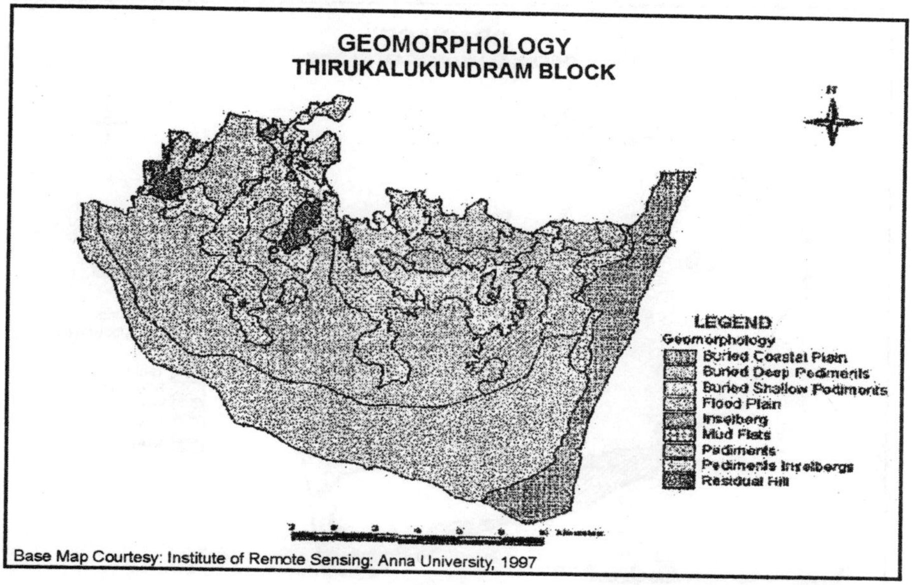

Fig. 16.2

Of the geomorphological divisions identified within the region, the Coastal Plain occupies the entire coastline of the block. Coastal Plains are formed due to long term wave action and is dominated by sandy features, and are not suited for locating aquafarms. So is the case with pediments, residual hills and inselbergs, which are essentially rocky and are not covered by sediments of any kind. Pediment may or may not have a thin veneer of alluvium over it. (Fig. 16.2)

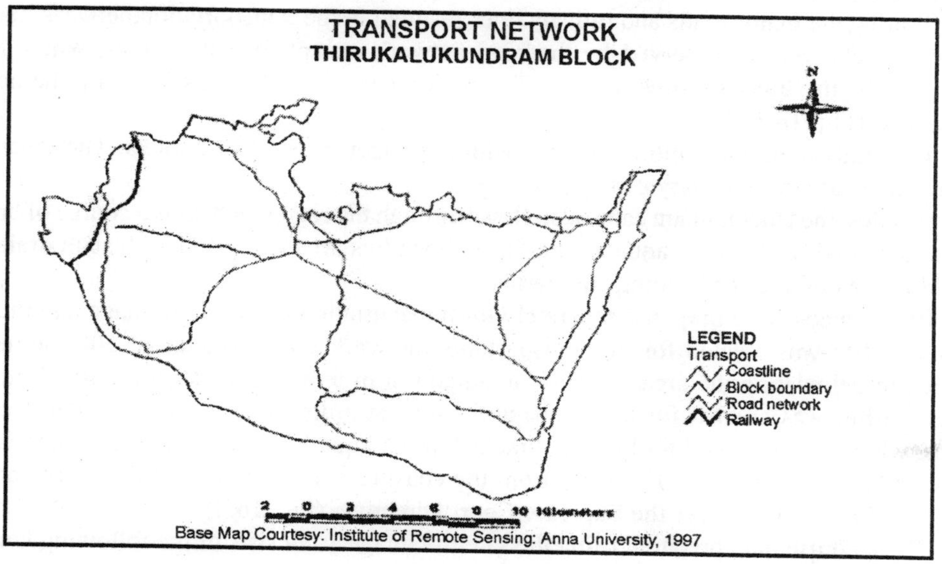

Fig. 16.3

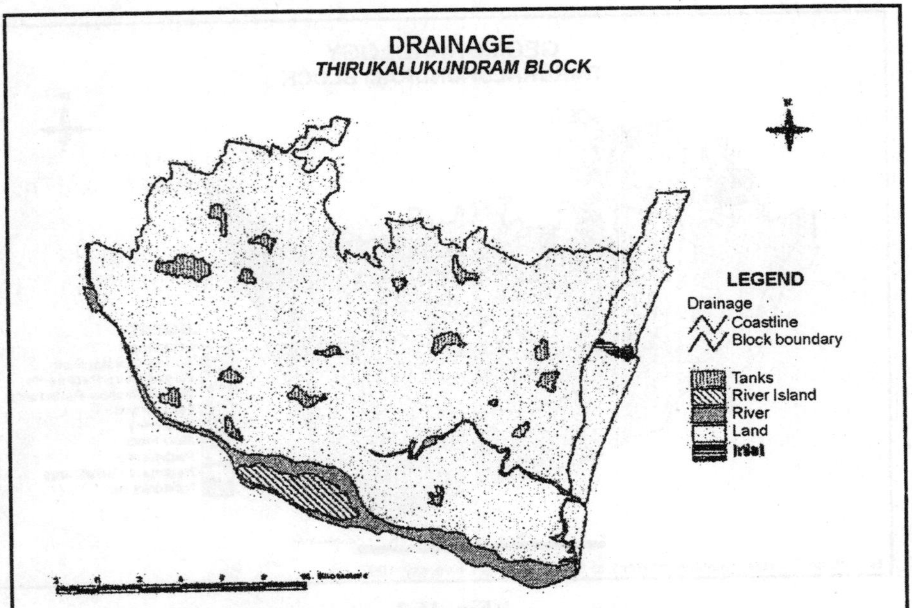

Base Map Courtesy: Institute of Remote Sensing: Anna University, 1997

Fig. 16.4

Mud flats characterised by clayey soils with low water permeability are coastal features formed due to the tidal action. Due to their impermeable nature, they are the most suited for locating aquaculture farms. The region also has a major flood plain along the banks of the Palar River, which flows along the southern boundary of the block.

Plotting the major roads and railway lines resulted in the transportation network map. The region is relatively well served by the road network except for some areas, which remain unserved by the transportation arteries. The region has a rail line passing along the northern boundary. (Fig. 16.3)

The study area is a region endowed with a number of tanks and inlets. The Palar River forms the southern boundary of the block.

Besides, the Buckingham canal also flows through the region, which is a source of brackish water essential for locating aquaculture farm. Since this area is blessed with such drainage, it could be favourable for locating aquafarms.

Buffer maps were prepared separately for transportation network, drainage and the coast. Single buffer was created for the transportation network for a distance of 500 metres. This helped in delineating the areas best for an aquafarm in terms of its accessibility. Similarly, a single buffer was created for the drainage features at a distance of 500 m to demarcate the areas, which are not suited for locating brackish water aquafarms. A buffer was also created for the coast at a distance of 300 metres from the coast to demarcate the CRZ, which cannot be used for aquaculture under the Supreme Court Directive. (Fig. 16.4)

The land use map prepared using the satellite imagery consists of the following land uses: (Fig. 16.5)

(1) *Cultivable land*—Most of the region falls under this land use category since most of the region is covered with buried pediments. Of the total area in the region 29.43 km^2 is used for agriculture.

(2) *Forests*—Forested lands dot the landscape in the northwest and they occupy an area of 20.09 km^2.

(3) *Waterbodies*—Numerous tanks are found in the region. The region is also blessed with a number of inlets. The Palar river forms the southern boundary of the block. The water bodies together occupy an area of around 15 km^2.

(4) *Built up area*—The region is almost rural with a dispersed settlement. The three major compact settlements are Thirukalukundram, Mahabalipuram, and Sadras. The built up area contributes 6.36 km^2.

(5) *Mudflats*—The most favourable land use for locating an aquafarm is none other than the mudflats. Mudflats are found on either side of the Buckingham canal.

(6) *Plantations*—The coast to the north of the block is dominated by Casuarina plantations. Further, other plantations are also found inland.

(7) *Sand Deposits*—Sand deposits could be found along the coast as well as along the river banks and this land use is not favourable for locating an aquaculture farm.

(8) *Dry lands*—A few patches of dry land can also be found within the region.

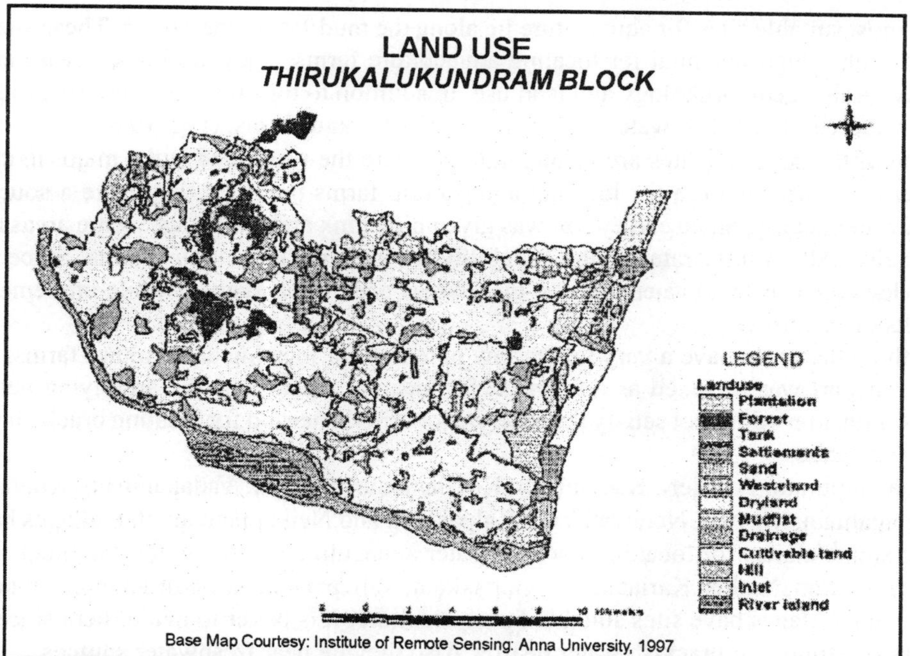

Fig. 16.5

The map obtained by overlaying the reclassed maps helped in identifying the potential sites for aquaculture. Areas with the highest aggregate weight are the most suitable sites for locating an aquafarm.

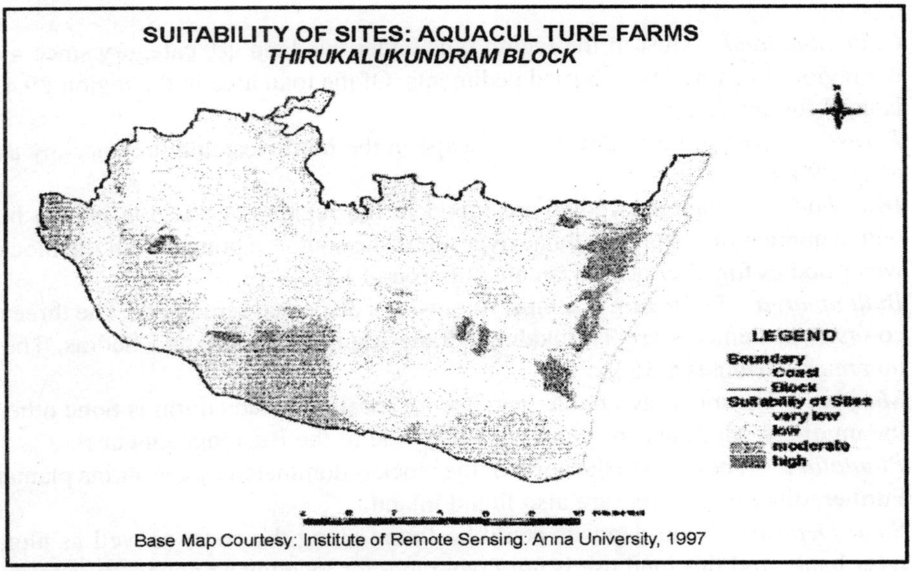

Fig. 16.6

The most suitable sites for aquaculture lie along the mudflats in the region. These sites are endowed with a high potential for locating aquaculture farms. They are blessed with clayey soil, a favourable geomorphology and land use. In addition to this, they have the Buckingham Canal—a source of brackish water which is essential for aquafarms. (Fig. 16.6)

Some of the potential sites are located inland where the wasteland is the major land use. Such areas can also be used for locating aquaculture farms provided they have a source of brackish water nearby. Since clayey soil was given high rank and weightage, some areas along the river also exhibit moderate to high suitability for locating aquaculture farms. Since these sites are located near freshwater sources, these areas need to be avoided while implementing the results of the project.

All the other areas have a very low to low potential for locating aquaculture farms. Even though transport was included as a criterion for locating aquafarms, the area lying near the transportation artery does not satisfy the other major criteria needed for locating brackish water aquaculture farms.

Mahabalipuram, Poonjeri, Kadambadi, Kokilamedur, Edaiyur, Vadakambadi, Kunnathur, Sthurangapatnam, Meyyur, Neduvakkarai, Kalpakkam and Neikuppam are the villages having the most suitable sites for locating brackish water aquaculrure farms within the block. Even though Virapuram, Nathan Kariacheri, Kiraipakkam, Vallapandai, Kolattur Paonpadapikuttam and Tirumani villages have sites suitable for aquaculture, it is better to avoid them since they do not have a source of brackish water nearby. Also they lie near freshwater sources.

CONCLUSION

From the study, it is evident that there is great potential for locating aquaculture farms along the coast of Thirukalukundram block. The coast is blessed with the right combination of soils, geomorphology and land use necessary for locating them. This study has also shown that area close to backwaters and areas lying within tidal flats are the most suited for brackish water aquaculture.

The results of this study clearly indicate the importance of GIS and remote sensing techniques in the present world scenario. It can be used to provide a variety of information needed for the development of aquaculture farms. Although this is a technically feasible approach, it needs to be answered whether it is cost-effective. Computer-automated method has helped in making the work time-effective as far as analysis and map updating are concerned. It also provides flexibility by allowing incorporation of new data. Modifications to existing data as well as manipulations can be done very quickly to produce maps, graphics or tables. The results are indicative of opportunities for aquaculture development and for planning. At the same time the present study has its own limitations in bringing down the environmental deterioration. It helps in locating the aquaculture farms in the most environmentally suited sites. But if the pollutants are let out from the aquaculture ponds without treatment, it will definitely polute the environment. GIS can be used further in designing techniques for waste disposal.

Investigations on the ground and in water are necessary to verify the suitability of the individual areas identified, before the results of this project are implemented. As in any other study, there is a certain amount of limitation as far as the accuracy is concerned. The maps have been obtained from different sources and the imagery for only one season has been used to derive the land use map.

In order to get better result, a number of other parameters need to be integrated for site selection of aquaculture farms. Inclusion of water quality information, socio-economic and environmental parameters in the analysis will certainly yield a more effective and worthy result. This study can be extended to other parts of the country and is an effective method for determining suitable sites for aquaculture development.

REFERENCES

Alagarswami, K., (1993), *Sustainable Management and Development of Coastal Ecosystem*, in M.S. Swaminathan and R. Ramesh (eds), Research Foundation and Institute for Ocean Management, Anna University, Madras.

Krishnamoorthy, R., (1993), Coastal Land Use Mapping for Brackish Water Aquaculture Site Selection using Satellite Data—A Case Study of Tamil Nadu Coast, *Proceedings of the National Symposium on Remote Sensing Application for Resources Management with Special Emphasis on Northeastern Region*, Guwahati: 215-227.

Ramachandran, S., (1993), Pilot Study on Coastal Zone Information System (CZIS) for Rameswaram Taluk, Interim Report, p. 35.

Ramesh, R. and Rajkumar, R., (1996), Coastal Aquaculture Site Selection and Planning in Tamil Nadu Using Remote Sensing and GIS, *Asian-Pacific Remote Sensing and GIS Journal*, 9(1): 39-49.

SECTION V

FORESTS AND THEIR MANAGEMENT

17

FOREST FIRE MANAGEMENT, GIS AND SUSTAINABLE DEVELOPMENT IN THE GARHWAL HIMALAYAS

D. MUKHOPADHYAY

INTRODUCTION

Forest fires cause enormous damage every year during April-July in different parts of India depleting precious forest cover, destroying a rich variety of flora and fauna including several species of rare plants and animals. A forest fire is an unenclosed and freely spreading combustion that consumes the natural fuels of a forest, *i.e.* duff, grass, weeds, bush, and trees. Forest fires occur in three principal forms, the distinctions depending essentially on their mode of spread and their position in relation to the ground surface. Surface fires burn surface litter and other loose debris of the first floor and small vegetation. Crown fires advance through the tops of trees or shrubs more or less independently of the surface fire and are the fastest spreading of all forest fires. Ground fires consume the organic material beneath the surface litter of the forest floor.

The causes of forest fire are numerous but they are primarily attributed to commission and omission by men. The villagers traditionally set some areas in the forest on fire in order to get fresh fodder for their cattle. These then spread to contiguous areas. According to forest officials, disgruntled poachers who feel hamstrung by the stepped-up anti-poaching measures also set forests on fire as a vindictive measure. Forest fires can also be caused when villagers make a bonfire in the farms to keep the wild animals away, or before collecting honey from a beehive. Sometimes forest fires are caused by natural factors. Intense heat wave conditions during peak of the summer season result in raging forest fire causing extensive damage. Fires also start spontaneously says from lightning by the rubbing of together bamboos during a storm. Sparks from coal-fired railway locomotives may also cause forest fires.

In some countries the setting of fires for agricultural cropland is an integral technique of agriculture. In other areas forest fire prevention, including public education, hazard reduction, and low enforcement, consumes a considerable amount of time and money. The two basic steps in preventing forest fires are reducing risk and reducing hazard. Hazard is reduced by compartmentalizing a forest with firebreaks and reducing the buildup of fuel by controlled burning.

Frequent fires occur in Garhwal and Kumaon regions during summer months when mean atmospheric temperature exceeds the normal level. However, the extension of such a dry period varies form place to place in different zones of Garhwal Himalaya. Some areas like Deoprayag, Srinagar, Tehri and Kotdwara (all < 100 m) are characterized by more prolonged dry spring, summer and winter periods compared to Uttarkashi, Pauri and Karanprayag (all between 900-1700 m). It is a well known fact that as one moves in the south of the Himalaya, the degree of prepitation decreases.

FOREST FIRES IN INDIA

In Himachal Pradesh over 450 cases of forest fires were reported in 1995. In addition to accidental wildfires, controlled burning had been used traditionally for forest clearances by villages for various purposes and as a means of management to promote the regeneration of desired tree species or the re-growth of grazing plants including grasses. Controlled burning has been used as a genuine forest management measure in the US for decades. In south and south-east Asia including India, "Slash and Burn" method of farming is used by the tribals of hilly areas, in which they cut down and burn small areas of the forest and use the cleared land for cultivation. This method of burning offers them not only the cheapest means to clear the forest, but also free fertilizers in the form of ash from the burnt vegetation on a limited scale. But when it is indiscriminately practiced, as is being done at present, the damage can be irreversible.

According to a study undertaken by Indian Space Research Organisation (ISRO), it was found that forest fires affected 21.5 per cent of the total forest area in the four hill districts of Uttaranchal i.e. Almora, Chamoli, Tehri and Pauri. The total fire affected area was 2,11,500 ha. Table 2.10 shows instances of fire in different years in Uttaranchal Hills.

Vulnerability

The most vulnerable stretches of the world are the youngest mountain ranges of Himalayas. The forests of Western Himalayas are more frequently vulnerable to forest fires as compared to those in Eastern Himalayas. According to reports available the devastation caused by fire in 1999 was very extensive with more than 80,000 hectares of forest turned to ashes by the first week of May. In 1995 fires in the U.P. hill had destroyed more than 3,75,000 hectares of forest wealth.

Damage due to forest fire are as follows:

- Damage to valuable forest produce especially timber and other minor forest produces.
- Loss of natural bio-diversity of a region.
- Loss to wildlife and its habitat.
- Loss in forest covers.
- Damage to natural regeneration in the form of seeding.
- Environmental degradation due to loss of carbon sink resource and increases in percentage of Carbon-di-oxide in atmosphere.
- Global change in microclimate and making environment unsuitable for living organisms.

- Damaging soil fertility and making it more prone to erosion.
- Depletion of ozone layer making life unhealthy.
- Heavy smoke causing breathing problem, and other related diseases.
- Loss in agricultural production due to soil erosion and change in climate.
- Loss of livelihood resources for the people depending on forests.

Ecological, economic and social impacts of the forest in India may be enlisted as:

Loss of timber, loss of bio-diversity, loss of wildlife habitat, global warming, soil erosion, loss of fuel wood and fodder, damage to water and other natural resources, loss of NTFP's. As per, rough estimate average tangible annual loss due to forest fire in the country is Rs. 440 crores (US $ 100 millions approximately).

The coniferous forest in the Himalayan region comprising of Firs (Albies spp.), Spruce (Picea smithiana), Cedurus deodara, Pinus roxburgli and P. wallichaiana etc. is very prone to fire. In Himalayan region, fires are common in May and June.

From 1995 to 1999, fire hazards assumed dangerous dimension. An area of 6,77,700 hectares was affected by these fires. The estimated timber loss from these hazards was US $43 millions. Besides, the other losses due to these fires were – loss of soil fertility, soil erosion, loss of employment, drying up of water resources and loss of bio-diversity. This fire brought a major change in the microclimate of the region in the form of soil moisture balance and increased evaporation. The thick smoke due to fire spread and affected visibility up to 14,000 feet.

Besides these major forest fires, the loss from the other fires reported from 13 states of the country for the period 1994-96 comes to us $ 20 millions. One other major fire reported from a sandal wood forest in Tamil Nadu for the year 1996-97 caused a loss of US $ 43 millions approximately.

Remote sensing and GIS can be used to find out the capital value of the loss in the study area and (i) susceptibility of fire (ii) estimation of environment related disasters. Clear understanding of the forest regarding their ecological and economic potential has been made possible using remote sensing and geographic information system. Monitoring the degradation of forest in terms of their production and protective functions and subsequent endangering of the habit at is possible. Forestry department and the remote sensing centre may work with special references to the use of satellite remote sensing and GIS in the detection and estimation of incident and extent of forest fire.

18

DETECTING CHANGE IN FOREST COVER IN SHIMLA DISTRICT USING REMOTE SENSING AND GEOGRAPHIC INFORMATION SYSTEM

PRAVEEN MISHRA

INTRODUCTION

To meet their basic requirements, human beings have been harnessing natural resources since they appeared on the earth. The phenomenal increase in population of both man and animal added with growing industralization and urbanization in the past few decades has led to faster degradation of natural resources. Scientists believe that in the ancient times 50 per cent of the earth's surface was covered with forest but now the forests cover only 30 per cent of the earth's surface. However, this percentage varies in different countries. The situation of India in this regard is gloomy i.e. on an average per person availability was 0.11 hectare in 1951, which has significantly decreased to 0.08 hectare in 2001. Similarly there is substantial difference between figure of forest cover as portrayed by Forest Survey of India and Satellite Imagery given by National Remote Sensing Agency, Hyderabad i.e. former claims there is 19.5 per cent forest cover in India while latter exhibit there is not more than 14 per cent forest cover (officially not announced). Therefore, survey gap of 6 per cent is very significant, which needs to be looked into. Various schemes have been launched for forest development during last four decades, but the impact of these schemes is not visibly reflected in the improvement of forest cover.

STUDY REGION

The study area is situated between 30°48′-31° 42′ north latitude and 77°-78° 15′ east longitude, at a mean altitude of 8,500 feet. This region is exactly on the watershed of two great river systems, the Sutluj and the Indus. This region is comprised of loose sedimentary rocks which are vulnerable to soil erosion. Shimla district lies in the lap of Himalayas. Its climate is largely conditioned by that single factor. In Shimla at about 7000 feet, the average temperature of coldest month is 4.7°C and of June, the warmest month, 19.5°C. While average annual temperature is around 12.7° C. At present the district supports a population of about 7.58 lakhs, with an average density of 145 person per square kilometre. The population growth rate of the district is 57.32 per cent. The year 1921 is considered as the watershed of the population growth in the district. Since then it is increasing at a rate of more than 20 per cent per decade.

The district is a predominantly agricultural one, where more than 70 per cent population is engaged in agricultural activities. Total forest area of the district is 2621 square kilometre, while the net sown area is 72.7000 hectare.

DATA SOURCE AND METHODOLOGY

The discussion is based on secondary data. It is obtained from various published and unpublished records of Government and voluntary organizations. The landuse and forest related data has been obtained from: (a) National Atlas and Thematic Mapping Organisation; (b) Forest Survey of India, toposheet no. 53 e,f,I; (c) Indian Remote Sensing, IC Satellite Imagery, 1999. Demographic data has been obtained from: (a) District Census Handbook; (b) India 2002; (c) Statistical Outline Himachal Pradesh.

Flow chart for detecting Change in Forest Cover in Shimla District

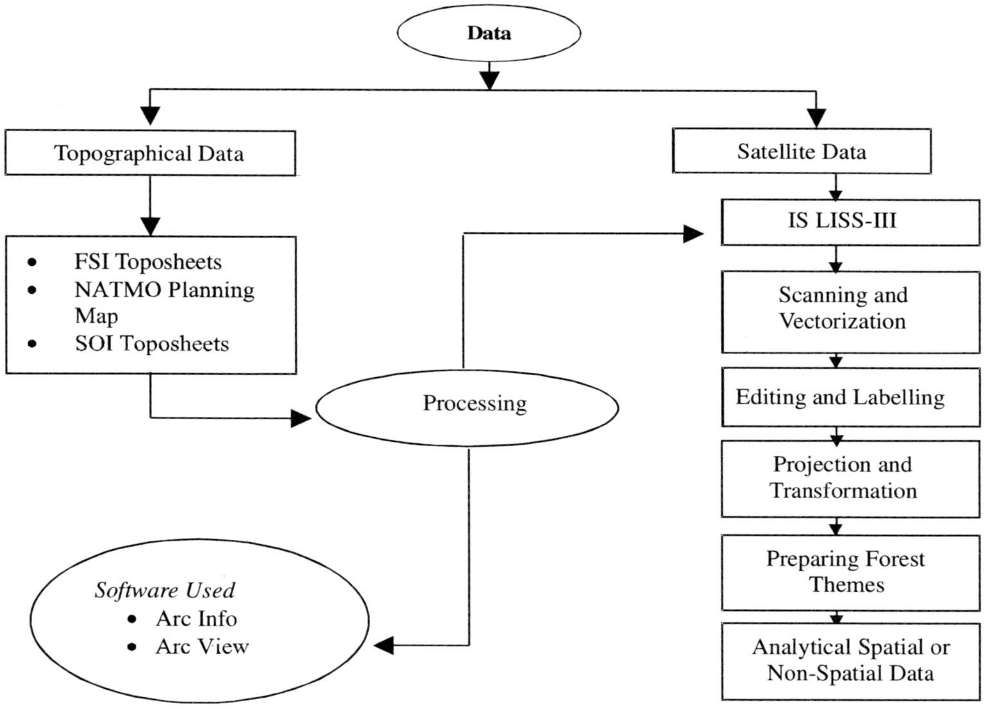

DISCUSSION AND RESULTS

There is sizable change in forest types in the period considered in our study, i.e. 1993 to 1999. In general total forest cover of Shimla District has decreased from 54 per cent to 52 per cent. Though being a hill district, it should be more than 60 per cent. Substantial proportion of negative change has been experienced in dense forest cover, which has decreased from 38 per cent to 31 per cent. On the contrary, open and scrub forest cover has increased from 7 per cent to 10 per cent and 8 per cent to 11 per cent respectively.

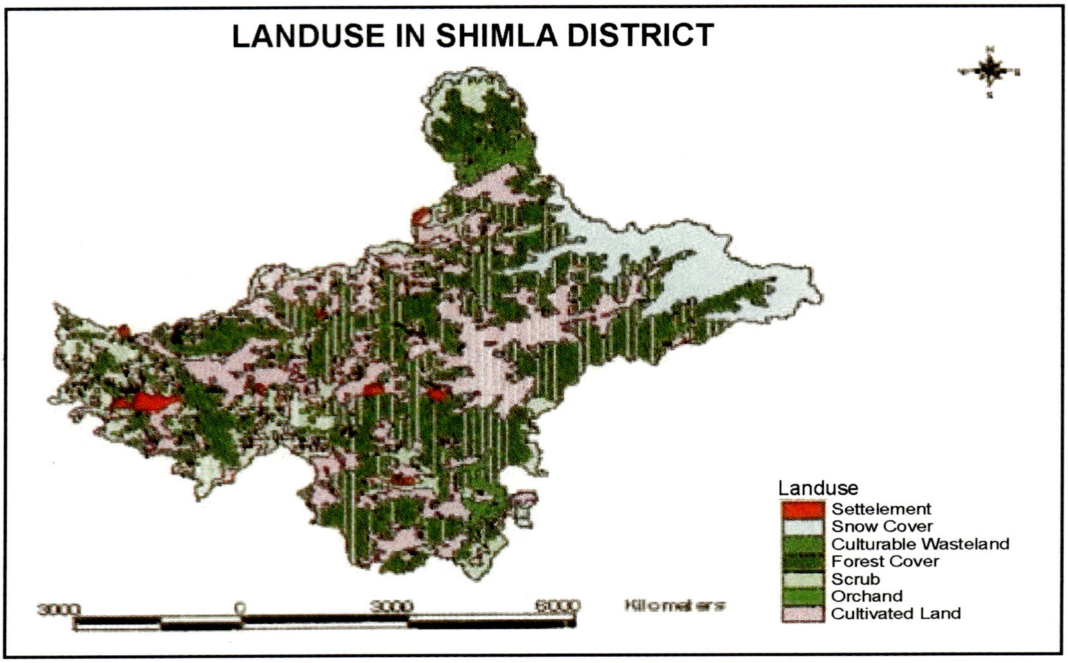

Map 18.1: Landuse in Shimla

Map 18.2: Forest Type in Shimla, 1993

Map 18.3: Forest Type in Shimla District, 1999

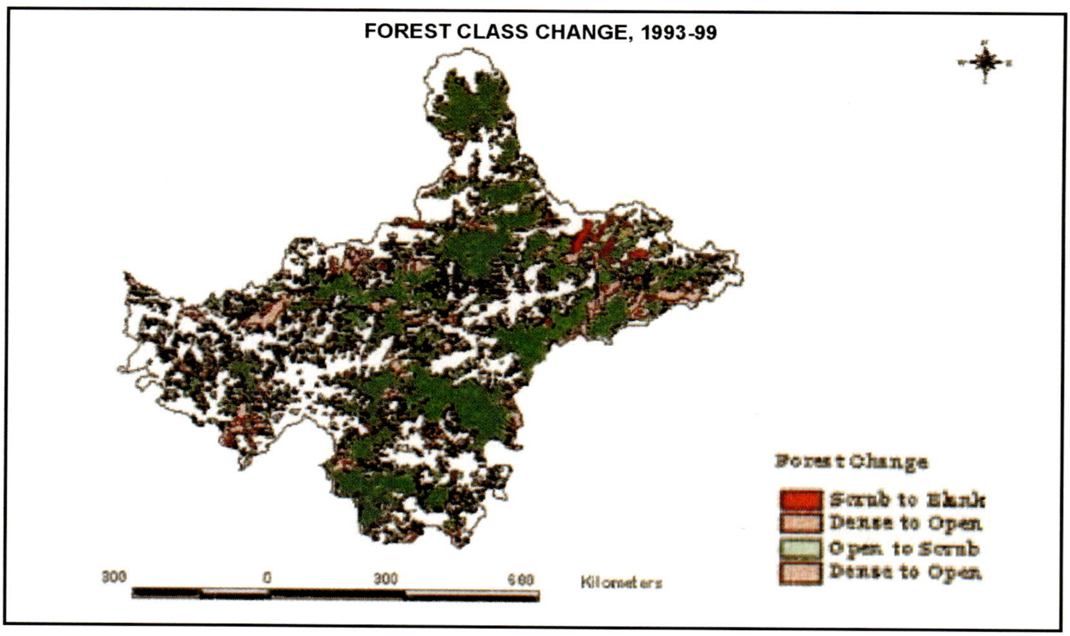

Map 18.4: Forest Change in Shimla District, 1993 to 1999

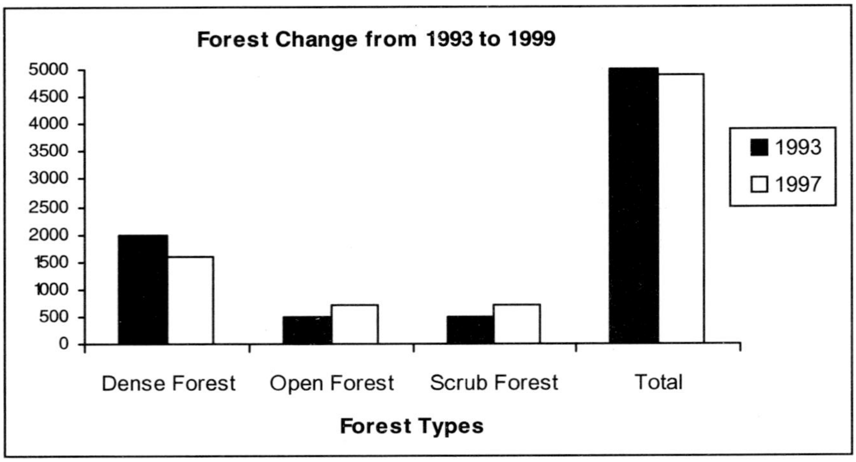

Fig. 18.1: Forest Change

It must be noted that increase in Open and Scrub forest cover is at the expense of decrease in Dense forest cover. The decrease in Dense forest cover is also ascribed to increase in area under habitation and arable land, as it is shown by IRS Satellite Imagery of the year 1999. Similarly, if we superimpose major towns of Shimla district for example Shimla Town, Rampur, Khashdhar, Rohru and Narkanda, it becomes evident that deforestation is mainly pronounced in the around the aforesaid mentioned town.

Table 18.1: Urban Population Growth in Shimla District

District	*Population Growth Rate (in per cent)*
Bilaspur	44.47
Chamba	40.36
Hamirpur	43.26
Kangra	21.27
Kinnaur	–
Kullu	24.15
Lahul Spiti	–
Shimla	57.32
Sirmaur	41.90
Solan	49.93
Una	31.69
State average	37.80

Source: Statistical Abstract of Himachal Pradesh, 1998

Further, if we observe the urban population growth rate of the Shimla district in comparison to remaining district of the Himachal Pradesh, it becomes obvious that it ranks Ist in terms of urban population growth rate. For example, it was 57.32 per cent during 1981-1991, and it is likely that this rate would continue in future because of high potential of urbanization in the district.

Table 18.2: Urban Expansion in Shimla District

Year	*Area in Sq.*	*Residential*	*No. of Households*
1971	19.55	9575	14731
1991	34.80	20059	36217

Bln = Blank, Scr = Scrub, Den = Dense, N/C = No Change, Ope–Open

The population of the Shimla district is increasing with frightening rapidity. (See Fig. 18.2), Consequently, the per capita forest cover of the Shimla district has decreased from 0.41 hectare to 0.34 hectare from 1993 to 1999. In Shimla district more than 70 per cent of the population resides in rural areas and they fulfil their energy requirements from fuel wood (88.35 per cent in rural and 5.35 per cent in urban area), which is obtained from nearby forest areas. Therefore, forest of the Shimla district is facing threat from both rural and urban area, but with specific reasons i.e. energy requirement and land requirement at the cost of forest and its area.

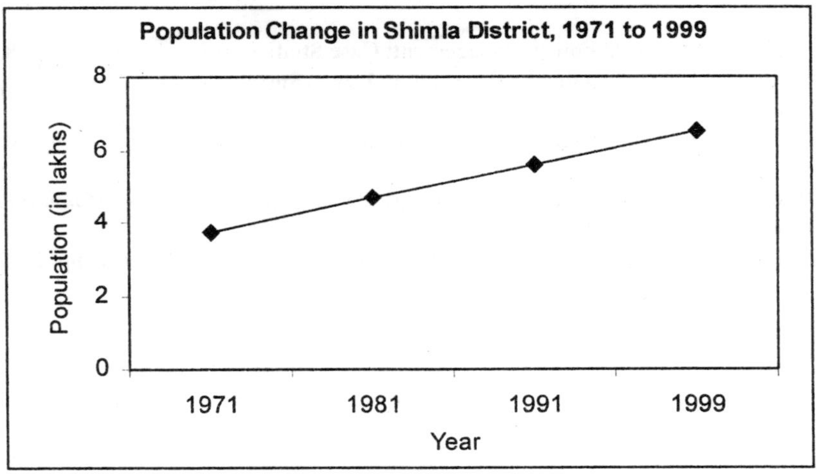

Fig. 18.2

Map 4 reveals that much forest has been degraded in the around Shimla town. It is obvious that Shimla town is important tourist place, and tourism involves certain activities which are detrimental to forest cover. For example, the development of hotels, restaurants, shops, picnic spots require a lot of place. That is why much clearing of forest cover has been taken in an around this famous tourist spot i.e. Shimla Town.

Further if we observe the north and central part of the district we find much forest clearing has taken place in this region. As we known that this part of the district is dominated by tribal population, and they still depend on the slash and burn agriculture, which is harmful to forest cover as well as fragile ecology of this region.

CONCLUSION

In our analysis it has become conspicuous that forest of the Shimla district requires immediate attention. The growing population and urban expansion beyond the carrying capacity of the region is challenging the fragile ecology of the region, which is already facing threat from the rural population's energy requirement.

REFERENCES

Dogra, B., May (2002), *Attempting Harmony of Livelihood and Conservation in Forests*, Yojana, Vol. 46, pp. 44-45.

Gade, D.W. (1996), *Deforestation and its Effects in Highland Madagascar*, Mountain Research and Development. Vol. 16, No. 2, pp. 101-116.

Ives, J.D. (1987), *The Theory of Himalayan Environmental Degradation: Its Validity and Application Challenged by Recent Researches and Application Mountain Research and Development*. Vol. 7, No.3.

Kumar, R., Rawat, K.N., August (2002), *Peoples Approach in Forest Management*, Kurukshetra, Vol. 50, No. 10, pp. 44-45.

Kaushal, K.K., May (2002), Forest *Management Technologies in Rural India*, Kurukshetra, Vol. 50, No. 7, pp. 43-45.

Mitra, A., August (1997), Joint Forest Management: Case Studies, *Yojana*, Vol. 41: No. 8, pp. 41-45.

Oberai, C.P., August (1997), Sustained Afforestation: Key to Sustainable Development, *Yojana*, Vol. 41, No. 8, pp. 31-33.

Panigrahi, S.K., January (2003), Environment Protection and the Tenth Plan, *Yojana*, Vol. 47, No. 1, pp. 37-42.

Price, M. (1987), Tourism and Forestry in Swiss Alps; Parasitism or Symbiosis? *Mountain Research and Development,* Vol. 7, No. 1, pp. 1-12.

Srinivasam, G., January (2003); Goals for the 10th plan in *Yojana*, Vol. 47, No.1, pp. 23-26.

19

CONSERVATION AND MANAGEMENT OF FOREST RESOURCE OF SIKKIM

RASHMI PRAKASH

INTRODUCTION

Forest is a renewable natural resource. It is considered a flow resource, which is very vital for human existence on the earth. For maintaining environmental balance and sustainable ecological conditions at least a 33 per cent vegetal cover of total land area is considered desirable.

Though natural resources are free gifts of nature to man, their conservation and management is essential so that they can be used for a longer period with maximum efficiency. Proper classification of natural resources is also needed for their proper management and conservation. This is because the same resource management principles cannot be applicable for all resources.

Forests, when properly managed, yield countless products, but when abused by over-cutting or poor management can become 'choked' flow resource. Forests are also dependent upon environmental conditions. Unfavourable environmental conditions diminish production from this resource. Therefore, for such resources it is more important to concentrate on managing the environment than the resource itself. A forest is a living and dynamic resource and its continuous harvest is dependent upon proper planning and management. Improper use may result in ecological degradation with adverse socio-economic consequences. Without proper management, conservation of forest is also not possible. Management and conservation can be possible by bringing in a sense of responsibility among the people, by government policies and control, by proper inventory and future projection of resources for which spatial information system is now of great help.

With the Remote Sensing Technology it is possible to observe very large areas of the earth and record data for various studies with the help of satellite borne cameras or sensors. This technique has helped in the collection of various types of data of immense importance. The imageries taken enable intensive mapping of the natural resources surveyed.

The same areas are observed at periodic intervals. Therefore, it is possible to monitor the changes occurring in the area of coverage and to plan for its management. It is with this perspective that forest cover can be observed over a large area and the changes occurring to them can be assessed and analysed.

In the absence of near real-time forestry information, gradual reduction in the forest wealth due to its continuous exploitation was not realized immediately. Various people and the different government departments gave varying figures and it was not possible to test their authenticity. This became possible with the launching of the Landsat-1, a series of satellites with a multi-spectral scanner onboard. At the same time different interpretation techniques were also developed with the aid of computers as well as visual means. By the early eighties remote-sensing scientists came forward to undertake operational investigation projects using remotely sensed data and produced highly dependable results. The scientists began to use this technique for delineating forest cover and for assessing the ecological status of the forests.

The National Remote Sensing Agency was established in India in 1975. It made a study of the forest surveys to compile gross forest vegetation resources information of the entire country on a single format and monitor the change. This established the credibility of remote sensing technique for assessing forest resources. It consisted of making rapid assessment of forest cover as it existed in 1972-75 and again in 1981-82, especially to demonstrate the deteriorating condition of the forests, and the use of this technique for rapid assessment. The data obtained from Landsat satellites revealed a reduction in the forest cover in India between 1972 and 1983. The country had lost about 2.8 per cent of the forest cover in about 7 or 8 years time.

To protect the forest cover the government started the Department of Forest and Environment. The Forest Survey of India (FSI) was established in 1975. FSI also adopted remote sensing for collecting information on forest wealth and for regular monitoring of forests-in every two years. Now, India has world's best satellite remotely sensed data from the remote sensing satellite series-IRS-IA, IRS-IB, IRS-IC, IRS-ID and P2 and P3 satellites. There are four regional offices of FSI-at Bangalore, Simla, Nagpur and Kolkata. Data of 1980-82 for the whole country was used and imageries were interpreted to delineate gross, dense, and open forest cover and maps were prepared. There was an increase of 2200 ha. in the forest cover between 1991 and 1993 in the country (FSI, 1993). This can be considered authentic. For the period 1993-95 the forest cover of the country was estimated at 19.27 per cent (dense forest 11.17 per cent, open forest 7.95 per cent, mangrove forest 0.15 per cent and scrubl. 74 per cent) of the total geographical area (India 2001), which was 23 per cent in 1974 (Central Forestry Commission). There can be many factors responsible for the change but the technique to monitor the change, the rate of change, the area covered, the types of species etc. is now reliable and very useful.

Some of the major information needed for managing forests are: dynamic changes and surveillance of forest cover, forest type distribution, plantation monitoring, forest inventory and volume estimation, yield calculation, preparation of zones or treatment areas, forest damage assessment, assessment of ecological and biodiversity related issues etc. These are now easily available with the help of this technology and these information can be effectively used for the conservation and management of forest resource.

ANALYSIS: FOREST RESOURCE OF SIKKIM

Sikkim, a tiny state in the eastern Himalaya, is bounded by China in the north, Bhutan in the east, Nepal in the west and by the state of West Bengal in the south. The state is richly endowed with forest wealth. Forests are one of the greatest assets of the state. In a hill state forestry holds a place of eminence as it provides valuable raw materials to some industries, offer numerous opportunities for economic development and also adds to the revenue of the state. Conservation and management of forest resource is very important for the maintenance of the ecological balance and sustainable environmental conditions of the state. That is why, in order to exploit forest wealth adequately a comprehensive plan of utilization linked with regeneration has been taken up.

Forests, to a great extent, depend upon the temperature, rainfall and physiography of the area. Sikkim, though a tiny state, has a varied natural vegetation cover. Sikkim is essentially a mountainous state with deep valleys, hills and ridges. It is totally devoid of flat land extensive enough to be called a plain. Mountains rise in elevation northwards and the general slope of the land is to the south. The altitude ranges from 300 m (lowest at the junction of the Tista and Ranjeet) to about 8583 m. According to altitude the land consists of lower and mid hills (300 m to 2000 m), Northern higher hills (2000 m to 5000 m), Alpine zone (above 5000 m), Snow covered tract (very high regions) and the valleys. The northern part is deeply cut into steep escarpments and except in Lachen and Lachung valleys, is almost uninhabited. Valleys are at elevations from 300 m to 4000 m. Southern Sikkim is of lower altitude with height ranging up to about 2440 m. The valleys here are comparatively open. Nearly 2/3 of Sikkim consists of perpetually snow-covered mountains. The entire state is an enclosed basin between two parallel and deeply intersected transverse ridges-Singli la and Donkya la. Southern and central Sikkim is a complex area with forest covered ridges and hills and intervening valleys. Tista is the main river flowing from north to south. The climate in Sikkim is varied, largely due to variations in altitude and temperature. Average annual rainfall ranges between 210 mm to 2500 mm temperature in lower altitude varies between 4.5°C to 18.5°C. There is a contrast in temperature and rainfall between sheltered valleys, mountain tops and high ridges. Climatic elements are controlled by altitude and location with respect to summer monsoon winds. The low valley bottoms have tropical climate which extends up to 2000 m. At higher altitude (up to 3000 m) the climate is temperate and warm. Higher up, the temperature goes down till it reaches the snowline. Even in snow-covered area the climate varies between Arctic and Polar due to altitude. Thus, a variety of climatic types is observed in this small state. The state receives heavy to very heavy rainfall from monsoon winds in summer and there is rainfall even in winter months. Because of heavy rainfall one finds tall trees and rich undergrowth, flowering plants of various types, rhododendrons and orchids. The state has a very rich forest wealth.

Great variation in climatic conditions in the different parts of the state, leads to a large variety and a luxuriant growth of trees and plants. The state can be divided into three vegetation zones having different types of forests: (a) Tropical and sub-tropical forest zone (stretching form the lowest level up to 1700 m) (b) Temperate zone (1700 m to 4300 m) (c) Alpine zone (form 4300 m up to the snowline of 5000 m). Roughly up to 2000 m almost every suitable part has been cleared for cultivation and forests remain only in the rocky ravines and the steepest slopes where no crops can be grown.

In the upper tropical zone, sal forests and bamboo trees of different types are in abundance. Different species of oak, bamboo, forests of cherry, laurels, chestnut and a prodigious variety of ferns are also found in this zone. In sub-tropical areas rattan, treeferns, plantains, screwpines and other tropical plants are available in the valley areas. Undergrowth of bush vegetation and leeches are also found. This type of vegetation is confined to a very limited area in southern Sikkim.

The temperate zone falls in comparatively northern and higher areas. It is basically the oak forest region. Here undergrowth is less dense. Bamboo and tall grasses form the undergrowth. Shrubby vegetation is also found in this zone. Trees of chestnut, maple, magnolia, pine, treeferns are found in the hills and on valley slopes. These are evergreen mountain forests. The forest is dense up to a height of 3000m with closed canopy. Branches and trunks of trees are thickly covered by moss.

Sub-alpine zone is practically the region of the rhododendrons and pine forests. Near Chhangu lake dwarf rhododendrons are found. In the Alpine zone the trees are found only up to a height of about 4000m but in exposed situations it is limited to a lower height. Here Alpine flora of western Himalaya and Siberian Alpine are found. Different varieties of orchids and flowering plants are also found. On limestone outcrops, giant trees called Cypruses Totulosa are found. Mixed forests are found with Cyprus trees, orchids and rhododendrons. Rhododendrons grow in abundance. With rise in altitude trees become shorter.

Forests in Sikkim have been classified as Dense forest and Open forest. Out of the total forest cover of 3118 sq km (43.94 per cent of the total geographical area) 75.70 per cent (2363 sq km) is under Dense forest and only 24.21 per cent (755 sq km.) under open forest. Out of the total geographical area of the state (7096 sq km) 33.30 per cent (2363 sq km) is under Dense forest and 10.64 per cent (755 sq km) is under open forest (Table 19.1).

Table 19.1: Forest Cover in Sikkim (1999 Assessment)

Type	*Area (sq km)*	*Per cent to Total Geographic Area*	*Per cent to Total Forest Area*	*Net Change (1997 to 1999) sq km*
Dense forest	2363	33.30	75.70	–60
Open forest	755	10.64	24.21	+49
Scrub	386	–	–	–53
Non-forest	3592	50.62	–	+64
Total forest cover	318	43.94	–	–

Source: State of Forest Report, 1999 (FSI, India) p. 4 of 5.

Forests have also been grouped according to their composition as Coniferous, Sal, and others. Coniferous forest has larger area compared to Sal.

Forests can also be categorised on the basis of ownership as Reserved forests, Protected forests and forests owned by Monasteries. Most of the forests are owned by the government and are reserved forests. About 100 sq km of forest area is owned by monasteries. Only 400 sq km has been kept as protected forest.

There is one National park and five wildlife sanctuaries covering about 28.88 per cent of the geographical area. It ranks as the best covered state in respect to percentage of protected area.

Fifty per cent of the total forest area is merchantable, while other areas are not profitable as far as revenue earning is considered. But, from ecological considerations all forests are valuable. Firewood, charcoal, timber, cardamom, medicinal herbs etc are forests products.

At present, the forest cover of the state, based on satellite data of Nov. 1998, is 3,118 sq km which constitutes 43.945 of the geographical area. (State of Forest Report, 1999). It was only

37.34 per cent in 1981 (Sikkim Statistical Handbook of 1981) and 36 per cent in 1976. It was 44.01 per cent (3139 sq km) in 1997 assesement. Thus, there has been an increase in the area but there has been a net decrease in the forest cover since the last assessment made in 1997 (Table 19.1). District-wise assessment of the forest cover has also been done with the help of remote sensing technique. There are four districts in the state. The data reveals that in all districts there are open and dense forests and scrub (Table 19.2). According to 1999 assessment of largest fraction of geographical area under forest cover is in the South district (73.12) and lowest (64.01 per cent) in the North, though North district has maximum area and south district has the minimum area under dense forest, compared to all districts. The area under open forest is also large in North district.

Table 19.2: District-wise Forest Cover (1999 Assessment)

	District →	*North*	*South*	*East*	*West*	*Total*
Dense Forest	Total Area sq km	847	421	50	594	2363
	% of total geog. Area	20.04	56.13	52.16	50.94	33.30
	% of area to dense forest area	35.84	17.82	21.20	25.14	100.00
Open Forest	Total Area sq km	332	129	115	179	755
	% of total geog. Area	7.85	17.30	12.05	15.35	10.64
	% of area to open forest area	43.92	17.09	15.25	23.71	100.00
Total forest area sq km		1199	550	616	773	3118
% to total forest area		37.81	17.64	19.64	24.79	100.00

Source: State of Forest Report, 1999 (FSI-India) p. 4 of 5.

There has been much emphasis by the government on the development, conservation and scientific exploitation of forest wealth of the state for increase in revenue earning. Forest Divisions were already created in the North, West and East districts with headquarters at Mangan, Temi and Gangtok respectively at the time of the merger of Sikkim. Residential quarters for forest personnel were constructed. Protection work was taken up in different areas in Sikkim by the forest department. Extraction of logs was done in North district. In watershed areas emergency work to protect forests were approved at Lema in North Sikkim, Shotak, Tathangchen, Maney Khola, Lower Burtuk, Tadong, Rabdong, Rongi, Martam, Mazitar, Tirkutam and Jorethang. New trees were planted to protect against landslides, landslips and soil erosion. This was done with the intention of protecting the environment and at the same time increasing the area under the cover. Trees of ornamental and economic value, especially sal trees, were planted along roads. Trees were planted in erosion-prone and degraded areas in Melli and Sang to regenerate the forest resource in these areas and protect the land from further erosion and environmental degradation. Planting of species of trees saplings of economic value was taken up by the Forest Department in all the different divisions of the state. A target of covering 1000 acres was achieved by 1981. In 1981 another 600 acres was brought under forest cover. (Progress with Peace, Govt. of Sikkim 1981). Fire protection scheme, Cardamom scheme, Medicinal Plants scheme were also taken up. At Sarsansa a nursery of medicinal herbs was established where pecae, Ranoulfia, Serpentina etc. are grown. To protect tender plants from biotic interference, fencing of vulnerable points with barbed wires was carried out. Maintenance of old plantation clearing, weeding, tending and cultural operations of plants and filling up of open areas measuring up to 2500 acres were taken up. Pulpwood plantation was also carried out and in 1989-91 more than 700 acres were covered.

To cope with the demands for planting saplings, nurseries were created. By 1981 such nurseries covered more than 90 acres of land and a large number of plants were available from state nurseries only.

Under the social forestry scheme about 300 acres of plantation was raised in 1980 and another 300 acres in Kashmahal and Gaucharan forests in 1981. This was to enable the farmers to use the plants for fuel, fodder and agricultural implements. Social forestry scheme is included under the "Rehabilitation of Degraded forests". Under social forestry scheme, especially in North Sikkim, saplings of fast-growing timber, fuel and fodder species where planted. Fuelwood plantation was taken up in East District near Gangtok and along army and border roads. Development of forest and plantation of trees were also taken up in those reserved forests where trees were overmature or much exploited or the species were not of economic value. It was realized that forests in Sikkim had overmature dead, and decaying trees. Left to the mercy of nature, these forests would eventually go waste and barren due to decay and decomposition as over-mature trees cannot reproduce younger crops. So, such forests call for removal and replacement of trees with faster growing economical and soil binding species to boost up timber and firewood production and also to help in the stability of the soil. Such forests were selected and efforts were made to regenerate these forests.

About more than 2000 acres of area, denuded and damaged by landslides and erosion, was afforested. Soil conservation nurseries raised seedlings and vegetative propagation materials i.e. rooted cuttings, rhizomes by bamboos, Cintrolla and Agave for afforestation throughout the state. Plantation was done by various agencies (Table 19.3).

Table 19.3: Forest Plantation in Sikkim (1951-1999)

Period	*1951-80*	*1980-85*	*1985-90*	*1990-92*	*1992-97*	*1997-99*	*TOTAL*
Area ('000 ha)	3.90	14.45	30.05	11.16	41.56	18.11	119.23

Source: State of Forest Report, 1999 (FSI-India) p. 3 of 5.

Under the forest protection scheme demarcation of boundaries of Khasmahal and Gaucharan lands was completed. Cardamom plantation had been maintained, inventory survey of reserved and other forests were made. Department took steps to have more trained personnel. Remote sensing technique was adopted in getting information about forests and in preparing inventory of forest cover.

During the decades of 1970 and 1980, forests were developed but were also exploited to earn maximum revenue, to cater to the demand of land for settlement by increased population, for increase in buildings in urban centers and especially by illegal feeling of trees by smugglers. This led to a loss of tree covered area, especially in South and West districts. In the south-western Sikkim feeling of trees, especially by smugglers, and neglect by the forest department, resulted in drying of streams and water crisis. Realising the importance of conservation of forests, the government took the task of conservation and afforestation very seriously.

Under the afforestration programme the government is trying to spread awareness on the importance of forest resource, educate and involve masses, motivate people to participate in the afforestation and conservation process. The chief minister spoke of bringing "Harit Kranti" through afforestation, conservation, and ban on grazing.

Cattle rearing is one of the major occupations of many communities in Sikkim. Extensive rearing of livestock is done especially by communities living in the higher altitudes and grazing harms the growth of forest. Sikkim has witnessed a drain of biodiversity due to overgrazing. After 1994, the state government banned the grazing in forests as a measure towards conservation of forest resources. However, it has not yet been scientifically proved that overgrazing damages forests.

The government also introduced Joint Forest Management-cum-Integrated Watershed Management Programme in the state in 1998. It has generated great interest among the people because they are being involved in the planning process. Sikkim has 30 watersheds. Afforestation under Joint Forest Management is being done by involving local people, NGOs, panchayats and the forest department. About 2191 ha area is covered under the programme. The JFM programme is extended to Khasmahal and Gaucharan areas and degraded forest lands. (The State of Forest Report, 1999).

To protect the forest land, the rich bio-diversity, and species of the forests of North Sikkim, the government scrapped the army's proposal to construct a G-Firing range in this area in 1997.

Thus, there has been continuous effort by the government to keep Sikkim green, to conserve the forest wealth and manage it in the best possible way.

CONCLUSION

Because of the great effort towards afforestation, conservation and management by the government there has been an increase in the total forest cover in the state which has now reached a level of 43.9 per cent of the total geographical area.

The 1999 assessment data reveal that there has been an overall decrease of 60 sq km of dense forest because of degradation of 49 sq km to open forest and 11 sq km to non forest. According to 1997 assessment 2423 sq km area was under dense forest which is reduced to 2363 sq km. There has been an increase of 49 sq km of open forest area. There has also been a decrease in the area under scrub. In East Sikkim, loss of forest area, compared to 1997 assessment, is due to encroachment, because of rapid rise in population and rapid urbanisation in this district. The improvement is due to regeneration of Shorea Robusta in south Sikkim and plantation of cryptomeria japonica in west Sikkim (State of Forest Report, 1999).

Remote sensing has helped in many ways in the conservation and management of forests. It was the combined information derived from satellite images and aerial photography and ground information that made it possible to get a better understanding of the forests. It helps in taking faster decisions in various ways of forest management. Remote sensing data are adopted by the state forest departments for surveillance of forest area, plantation monitoring, forest inventory and preparation of treatment areas etc. It is easier and quicker to get various types of data related to forest at state and district level which can be used for effective management of forests of the state. Such spatial distribution data are necessary, especially to plan what type of species need to be introduced by replacing natural forest or for regenerating degraded forests in a more scientific manner. Besides, for stock mapping monitoring of deforestation, afforestation, encroachment area etc. are important. Various agencies, like ISRO, are obtaining data from different satellites in cooperation with forest administration. Effective forest management is now possible with the use of GIS Techniques by integrating information from all sources which could help in preparing databases and in updating data even at district or block level.

Thus, it is with a holistic approach, joint forest management programme and remotely sensed information, conservation and management of forest resource has been possible in Sikkim.

REFERENCES

Census of India 1981, Series 9-Sikkim, District Census Handbook, North, East South & West Districts, Directorate of Census Operations, Sikkim, Gangtok, 1984, 12-14.

Centre for Science and Environment, *Down to Earth,* Feb. 15 (1999) New Delhi, pp. 30-35.

Government of Sikkim, Budget 1980-1981, Budget Speech by Minister of Finance. Sikkim Government Press, Gangtok.

India 2001, Publications Divisions, Ministry of Information and Broadcasting, Government of India, New Delhi. pp. 191-92.

Jana, M.M. (2000), Causes of Depletion of Forests and their Conservation in India, *Indian Journal of Landscape System and Ecological Studies*, Vol. 23 No. II. Dec. 59-69.

Narayana, L.R.A. (1999), Remote Sensing and its Applications, University Press (India) Ltd. 3-5-819, Hyderabad.

Government of Sikkim, Gangtok. (1981), Progress with peace: A Review of Achievements of the present Government in the last months of its Administration, *Information and Public Relation Department,* 37-38.

Reddy M. Anji (2002), Textbook of Remote Sensing and Geographical Information Systems, Second Edition, B.S. Publications, Hyderabad. 2002.

Singh L.R., Singh, S., Tiwari, R.C., Srivastava. R.P. (Eds.) Environmental Management. The Allahabad Geographical Society, Department of Geography, University of Allahabad, Allahabad.

Forest Survey of India (1999), *The State of Forest Report*, Dehradun, Chapter III. 1-5.

INDEX